STUDY GUIDE AND

SOLUTIONS BOOK FOR

ORGANIC CHEMISTRY:

A SHORT COURSE

Fourth Edition

HOUGHTON MIFFLIN COMPANY

BOSTON

New York

Atlanta

Geneva, Ill.

Dallas

Palo Alto

STUDY GUIDE AND

SOLUTIONS BOOK FOR

ORGANIC CHEMISTRY:

A SHORT COURSE

Fourth Edition

HAROLD HART AND

ROBERT D. SCHUETZ

Michigan State University

ISBN: 0-395-04575-4

INTRODUCTION FOR THE STUDENT

This study guide and answer book was written for you, to help you master the subject matter of our text. The principles and facts of organic chemistry are not easily learned by simply reading them, even repeatedly. Formulas, equations, and the subtleties of molecular structures are best mastered by *written* practice. For this reason we have devised many questions at the end of each chapter in the text, to help you become thoroughly familiar with and understand the material in each chapter.

But it is our experience that such questions are not put to their best use unless correct answers are also available. Indeed, the answers alone are not enough. If you know how to work a problem, and find that your answer agrees with the correct one, fine; your ego is boosted, and you go on to the next problem with some confidence. But what if you work conscientiously and yet cannot solve the problem? You then succumb to temptation, look up the answer, and encounter another quandary--how in the world did the author get that answer?

This book has been designed with this difficulty in mind. For many of the problems, all the reasoning involved in getting the correct answer is spelled out in detail. Sometimes supplementary help is given, as for example the hints on writing electron-dot formulas included in the answer to Problem 1-4. Almost all of the answers to synthesis questions include cross-references to equations in the text. This should help you to see that the reaction which may be required to solve a particular problem is merely a specific example of a general equation given in the text. Such cross-references pertain not only to the chapter where the problem occurs but to equations in earlier chapters as well, thus helping you to review continuously. If you cannot solve a particular problem, these references will also guide you to parts of the text which you should review.

To help you locate the answer to a problem quickly, the pages in this book have been numbered at the top to correspond to the chapter and problem number in the text.

Finally, just a word of advice about how to master the many reactions you will study during this course. Some system is necessary, and we suggest the following approach. First, learn the nomenclature systems thoroughly for each new class of compounds you encounter. Then, rather than memorize the particular example which may be cited in the text, study reactions as typical of the class of compounds. Suppose, for example, that you are asked on an examination to write equations for the synthesis of 2-butanol? How to begin? Do not ask yourself: How do I synthesize 2-butanol? Instead, say: 2-butanol is an alcohol (*-ol* ending to the name). Then ask: what general ways do I know for preparing alcohols? Perhaps you write them out (using R for any organic radical, rather than a specific compound). Now ask: which of these methods can be applied to 2-butanol? Select the best, and use the general equation as a guide to construct the specific equation for 2-butanol. Gradually, with practice, short cuts to this procedure can be introduced. This approach will probably help you to eliminate a great deal of the memory work often associated with courses in organic chemistry. We urge you to study regularly; and we hope that this study guide–answer book will make it easier for you to do so.

A NOTE TO THE INSTRUCTOR

Swelling enrollments in our colleges and universities have had many consequences for both professor and student. One of these, an increase in student-to-faculty ratio, has necessarily resulted in a decline in the number of small recitation sections where individual student difficulties can be handled. A greater burden for self-education has been placed on the student's shoulders. To help ease this burden, we have written this book.

Great effort has been expended to ensure the accuracy of the answers in this book. It is easy, however, for errors to creep in, and we will be particularly grateful to those who call them to our attention. Suggestions for improving this book will also be welcomed.

CONTENTS

CHAPTER ONE

GENERAL PRINCIPLES

1-1. In sodium chloride, Cl is present as chloride ion (Cl^-); Cl^- reacts with Ag^+ to give AgCl, a white precipitate. The C—Cl bonds in CCl_4 are covalent; no Cl^- is present to react with the Ag^+.

1-2. Ionic compounds are formed from elements which differ widely in their electronegativity: NaF, $MgCl_2$, LiCl. Covalent compounds are formed from elements with identical or similar electro-negativities: F_2, P_2S_5, S_2Cl_2.

1-3. *a*) $:\ddot{C}l\!-\!\ddot{C}l:$ *c*) $:\overset{\longleftrightarrow}{\ddot{O}}\!=\!C\!=\!\overset{\longleftrightarrow}{\ddot{O}}:$ *e*)

(pure covalent)

b) $H\!-\!\overset{H}{\underset{H}{\overset{|}{C}}}\!-\!\ddot{F}:$ *d*) $H\!-\!\overset{\longrightarrow}{\ddot{B}r}:$ *f*) $H\!-\!\overset{H}{\underset{H}{\overset{|}{C}}}\!-\!H$

1-4. *a*) CH_3Cl: Carbon contributes four valence electrons, each hydrogen contributes one, and the chlorine contributes seven. Therefore, 14 valence electrons are available to bind the five atoms together. This must be done in such a way that the carbon and chlorine have eight electrons around them, and each hydrogen has two. Hence:

$$H:\overset{H}{\underset{H}{\ddot{C}}}:\ddot{C}l:$$

b) C_3H_8: There are 20 valence electrons (3 × 4 from carbon and 8 × 1 from hydrogen).

$$H:\overset{H}{\underset{H}{\ddot{C}}}:\overset{H}{\underset{H}{\ddot{C}}}:\overset{H}{\underset{H}{\ddot{C}}}:H$$

c) C_2H_5F: Again, 20 valence electrons. Halogens (F, Cl, Br, I) usually have three unshared electron pairs (shown on the following page).

1

$$\begin{matrix} & \text{H} & \text{H} & \\ & \overset{..}{} & \overset{..}{} & \\ \text{H} & \overset{..}{\underset{..}{C}} & \overset{..}{\underset{..}{C}} & \overset{..}{\underset{..}{F}} \\ & \text{H} & \text{H} & \end{matrix}$$

H H
H:C:C:F:
H H

d) HCN: Ten valence electrons (N is in the fifth group in the periodic table, and contributes five valence electrons). One has to know that the carbon is in the middle.

$$\text{H}:\text{C}:::\text{N}:$$

e) CO_3^{-2}: Twenty-four valence electrons (don't forget that this is a negative ion; two electrons come from some metal, such as sodium). The carbon is surrounded by the oxygens. The choice of which of the three oxygens is doubly-bound to carbon is purely arbitrary.

$$\left[\begin{matrix} & :\ddot{O}: & \\ :\ddot{O}: & \ddot{C}:: & \ddot{O}: \end{matrix} \right]^{-2}$$

In the formula as written above, the carbon and doubly-bound oxygen are formally neutral. The *carbon* is surrounded by eight electrons, but shares all of them with oxygens. Its share of these electrons is therefore four, which is exactly the number of valence electrons in a neutral carbon atom. The *doubly-bound oxygen* has two unshared electron pairs which therefore "belong" entirely to it. It also shares four electrons with carbon, its share being two. Thus it has a total of six valence electrons, which is precisely that of a neutral oxygen atom.

But the *singly-bound oxygens* "own" seven electrons each; they have six electrons in the three unshared pairs, and they also "own" one electron of the pair shared with carbon, for a total of *seven*. Neutral oxygen atoms, however, have *six* valence electrons. These oxygens therefore each carry a single negative charge. Since there are two of them, the entire species, carbonate ion, carries two negative charges which are "formally" located on the singly-bound oxygens.

Since one can write two additional electron-dot formulas for carbonate ion in which each oxygen in turn is doubly-bound to carbon, the average charge will be $-2/3$ on each oxygen.

f) NO_3^{-1}: Nitrogen contributes five electrons, each oxygen six, and one electron comes from the counter-cation, for a total of 24.

$$\left[\begin{matrix} & :\ddot{O}: & \\ :\ddot{O}: & \text{N}:: & \ddot{O}: \end{matrix} \right]^{-1}$$

This formula looks exactly like that for carbonate ion, just discussed above, except that the central atom is nitrogen, not carbon. The "formal" charges, calculated as above, are −1 for each of the singly-bound oxygens, 0 for the doubly-bound oxygen, and +1 for the nitrogen (a neutral nitrogen atom has five valence electrons, whereas the nitrogen in nitrate ion "owns" only four of the eight electrons it shares with the oxygens). The net charge on the ion is therefore −1 [2(−1) + 0 + 1 = −1].

Since the choice of doubly-bound oxygen is arbitrary and might be made in any of three ways, the average charge on each oxygen is −2/3, and the charge on nitrogen is +1.

g) CH_3NH_2: The order of attachment is as shown. There is one unshared electron pair remaining on the nitrogen, and there are no formal charges.

$$
\begin{array}{c}
\text{H} \\
\ddot{}\ \ \ddot{} \\
\text{H:}\underset{\cdot\cdot}{\overset{\cdot\cdot}{\text{C}}}\text{:}\ddot{\text{N}}\text{:H} \\
\text{H\ \ H}
\end{array}
$$

h) Cl_2:

$$:\ddot{\text{Cl}}:\ddot{\text{Cl}}:$$

i) C_2H_5OH:

$$
\begin{array}{c}
\text{H\ \ H} \\
\ddot{}\ \ \ddot{}\ \ \ddot{} \\
\text{H:}\underset{\cdot\cdot}{\text{C}}\text{:}\underset{\cdot\cdot}{\text{C}}\text{:}\underset{\cdot\cdot}{\text{O}}\text{:H} \\
\text{H\ \ H}
\end{array}
$$

Twenty electrons are available (eight from the two carbons, six from the hydrogens and six from the oxygen). The oxygen ends up with two unshared pairs, but there are no formal charges.

j) CH_3^+: Carbon contributes four electrons, and each hydrogen three, for a total of seven. But the fragment is positively charged, so one of these seven must have been donated to some other species, leaving six.

$$
\left[
\begin{array}{c}
\text{H} \\
\ddot{} \\
\text{H:}\underset{\cdot\cdot}{\text{C}} \\
\text{H}
\end{array}
\right]^{+1}
$$

The carbon carries a formal +1 charge. Since the octet around carbon is incomplete, the fragment is very reactive and combines with any reagent which can donate an electron pair to it.

1-5. For definitions and/or examples, see especially sections 1-2 and 1-6 of the text; also, consult the index.

1-6.

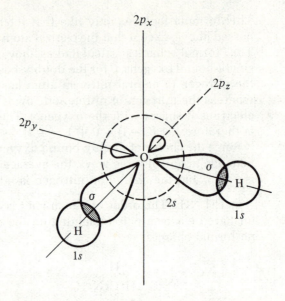

The predicted angle is 90°; the observed angle may be larger due to repulsion between like ends of the O—H dipoles.

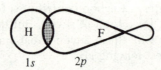

Here the plus ends of the dipoles are farther apart.

1-7. *a)* HF:

H ⬤ F
1s   2p

The molecule is diatomic, therefore necessarily linear.

b) H₂O₂:

H
|
O—O   or
      |
      H

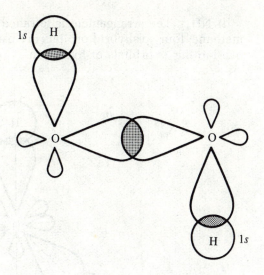

2p$_x$ and 2p$_y$ on the oxygen, 1s on the hydrogen

c) NH$_3$:

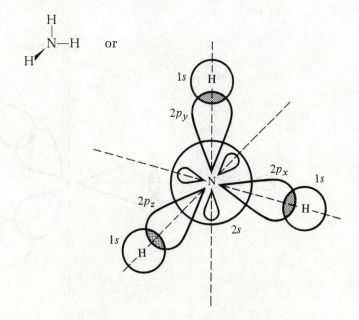

The unshared pair on N is placed in the 2s orbital (other formulations are possible).

d) NH_4^+: The arrangement is tetrahedral, entirely analogous to methane; four sp^3 hybrid orbitals are used for nitrogen, and they overlap the 1s orbitals of hydrogen.

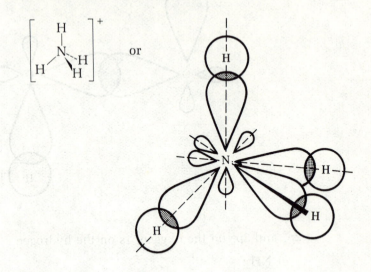

e) BF_3:

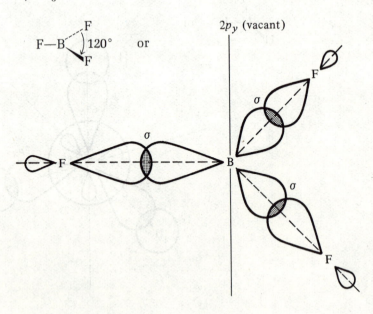

The three boron electrons are in three sp^2 orbitals. Each of these overlaps with a 2p orbital of fluorine.

1-8. The best arrangement is to have the three hydrogens lie in a plane, with H—C—H angles of 120°. The vacant orbital (+ charge) is a p-orbital perpendicular to this plane.

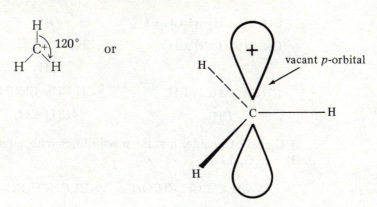

1-9. One hybridizes the $3s$ and three $3p$ orbitals of silicon to make four sp^3 hybrid orbitals, exactly as for carbon in methane. The geometry is tetrahedral.

1-10. *a*) C_4H_{10}: CH_3—CH_2—CH_2—CH_3 or CH_3—$\underset{\underset{CH_3}{|}}{CH}$—$CH_3$

b) C_2H_6O: CH_3—CH_2—O—H or CH_3—O—CH_3

c) C_3H_8O: CH_3—CH_2—CH_2—O—H or CH_3—$\underset{\underset{O-H}{|}}{CH}$—$CH_3$

or CH_3—O—CH_2—CH_3

d) C_2H_7N: CH_3—CH_2—NH_2 or CH_3—NH—CH_3

e) $C_2H_3Br_3$: $H-\overset{\overset{H}{|}}{\underset{\underset{H}{|}}{C}}-\overset{\overset{Br}{|}}{\underset{\underset{Br}{|}}{C}}-Br$ or $H-\overset{\overset{H}{|}}{\underset{\underset{Br}{|}}{C}}-\overset{\overset{Br}{|}}{\underset{\underset{H}{|}}{C}}-Br$

f) C_3H_6: CH_3—CH=CH_2 or $\underset{CH_2}{CH_2\text{—}CH_2}$ (cyclopropane)

g) C_3H_4: CH_3—C≡CH or CH_2=C=CH_2

or $\underset{CH_2}{HC\text{=}CH}$

1-11. a) C_6H_{14}: The problem should be approached systematically. Consider first a chain of six carbons, then a chain of five carbons with a one-carbon branch, etc.

$$CH_3CH_2CH_2CH_2CH_2CH_3$$

$$CH_3\underset{\underset{CH_3}{|}}{CH}CH_2CH_2CH_3$$

$$CH_3\underset{\underset{CH_3}{|}}{CH_2}CHCH_2CH_3$$

$$CH_3-\underset{\underset{CH_3}{|}}{\overset{\overset{CH_3}{|}}{C}}-CH_2-CH_3$$

$$CH_3\underset{\underset{CH_3}{|}}{CH}-\underset{\underset{CH_3}{|}}{CH}CH_3$$

b) $C_4H_{10}O$: Consider first the possibilities with a hydroxyl group (OH); i.e., C_4H_9OH.

$$CH_3CH_2CH_2CH_2OH$$

$$CH_3-\underset{\underset{CH_3}{|}}{CH}-CH_2OH$$

$$CH_3\underset{\underset{OH}{|}}{CH}CH_2CH_3$$

$$CH_3-\underset{\underset{CH_3}{|}}{\overset{\overset{OH}{|}}{C}}-CH_3$$

Then consider those with a C—O—C arrangement.

$$CH_3-O-CH_2CH_2CH_3$$

$$CH_3CH_2-O-CH_2CH_3$$

$$CH_3-O-\underset{\underset{CH_3}{\diagdown}}{\overset{\overset{CH_3}{\diagup}}{CH}}$$

c) C_4H_9Cl: Consider first compounds with a consecutive chain of four carbon atoms, then those with a branched chain.

$$CH_3CH_2CH_2CH_2Cl$$

$$CH_3-\underset{\underset{CH_3}{|}}{CH}-CH_2Cl$$

$$CH_3CH_2\underset{\underset{Cl}{|}}{CH}CH_3$$

$$CH_3-\underset{\underset{Cl}{|}}{\overset{\overset{CH_3}{|}}{C}}-CH_3$$

d) C_3H_9N: The carbon chain may be consecutive or it may be interrupted by nitrogen.

$$CH_3CH_2CH_2NH_2$$

$$CH_3-NH-CH_2CH_3$$

$$CH_3\underset{\underset{NH_2}{|}}{CH}CH_3$$

$$CH_3-\underset{\underset{CH_3}{|}}{N}-CH_3$$

1-12. *a)* C_3H_8: possible; see Problem 1-4.

b) C_4H_9: not possible; one cannot have an odd number of H's if all the other atoms present have *even* valences.

c) $C_2H_4Br_2$: halogens have the same valence as hydrogen, so this would correspond to the hydrocarbon C_2H_6 (ethane). Two structures are possible, depending on whether the bromines are on the same or different carbons; draw them.

d) C_2H_5Cl: possible; see preceding answer.

e) C_4H_{11}: too many H's for the number of C's; not possible.

f) CH_4O: one structure possible; the C and O are bonded.

1-13. Many answers are possible for each case; only one example is given. See if you can devise another.

a) CH_5N: CH_3—NH_2

b) CH_4S: CH_3—SH

c) C_3H_8O: $CH_3CH_2CH_2OH$

d) C_2H_5OCl: HO—CH_2—CH_2—Cl

1-14. 0.660 g of CO_2 is equivalent to $12/44 \times 0.660$ g of carbon or 0.180 g of carbon. The compound therefore contains $0.180/0.300 \times 100 = 60\%$ carbon.

Similarly, the sample contains 0.04 g of H or 13.3% H. By difference, it must contain 26.7% O. The empirical formula is then calculated as follows:

$$
\begin{array}{lll}
C & 60/12 = 5.00 & 3 \\
H & 13.3/1 = 13.3 \quad \text{or} & 8 \quad \text{or} \quad C_3H_8O \\
O & 26.7/16 = 1.67 & 1
\end{array}
$$

C_3H_8O is also the molecular formula (mol. wt. 60). For possible structures, see Solution 1-10.

1-15. 0.178 g of $CO_2 = 12/44 \times 0.178 = 0.0486$ g of carbon; $\%C = 0.0486/0.250 \times 100 = 19.4\%$.

0.146 g of $H_2O = 2/18 \times 0.146 = 0.0162$ g of hydrogen; $\%H = 0.0162/0.250 \times 100 = 6.48\%$.

41.8 ml $\times$ 0.200 N HCl = 8.36 milliequivalents of HCl = 8.36 milliequivalents of NH_3 = 8.36 milliequivalents of N = 0.117 g of N; $\%N = 0.117/0.250 \times 100 = 46.8\%$. By difference, the substance contains 27.3% O. The empirical formula is then calculated as follows:

$$
\begin{array}{lll}
C & 19.4/12 = 1.618 & 1.0 \\
H & 6.48/1 = 6.48 \quad \text{or} & 4.0 \\
N & 46.8/14 = 3.34 & 2.06 \\
O & 27.3/16 = 1.70 & 1.05
\end{array}
$$

The best molecular formula is CH_4N_2O; the substance is urea:

$$H_2N-\overset{\overset{\textstyle O}{\|}}{C}-NH_2$$

1-16. *a, c, e,* and *i:* all have OH groups (alcohols). *b* and *g:* both have two carbon atoms attached to oxygen (ethers). *f* and *h:* both have carbon-bound halogen atoms. The structural formulas (in abbreviated form) are given in the problem for *a, b, f, g,* and *h.* The others are:

c) $\underset{\underset{\textstyle OH}{|}}{CH_2}-\underset{\underset{\textstyle OH}{|}}{CH}-\underset{\underset{\textstyle OH}{|}}{CH_2}$

e) $CH_3CH_2CH_2CH_2OH$ and three isomers

d) $CH_3CH_2CH_2CH_2CH_3$ and two isomers

i) $CH_3CH_2CH_2OH$ and one isomer

CHAPTER TWO

SATURATED HYDROCARBONS

2-1. *a*) 3-Methylpentane: first, locate the root of the name (in this case, *pent*), write down and number the carbon chain.

$$\overset{1}{C}-\overset{2}{C}-\overset{3}{C}-\overset{4}{C}-\overset{5}{C}$$

Next, locate the substituents (3-methyl)

$$\overset{1}{C}-\overset{2}{C}-\underset{\underset{\textstyle CH_3}{|}}{\overset{3}{C}}-\overset{4}{C}-\overset{5}{C}$$

Finally, fill in the remaining hydrogens.

$$CH_3-CH_2-\underset{\underset{\textstyle CH_3}{|}}{CH}-CH_2-CH_3$$

b) $CH_3-\underset{\underset{\textstyle CH_3}{|}}{CH}-\underset{\underset{\textstyle CH_3}{|}}{CH}-CH_3$

c)
$$CH_3-CH_2-\underset{\underset{CH_3}{|}}{\overset{\overset{CH_3}{|}}{C}}-\underset{\underset{CH_2CH_3}{|}}{CH}-CH_2-CH_3$$

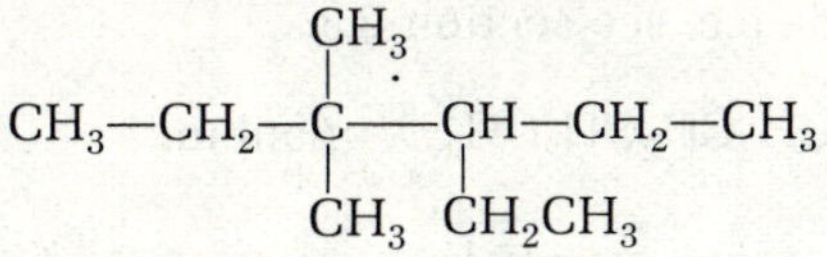

d)
$$CH_3-\underset{\underset{Cl}{|}}{CH}-\underset{\underset{CH_3}{|}}{CH}-CH_2-CH_3$$

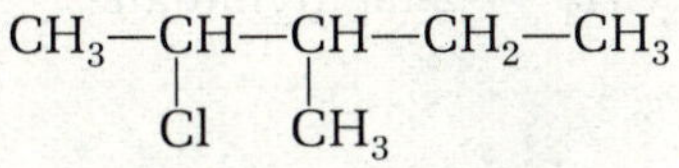

e) 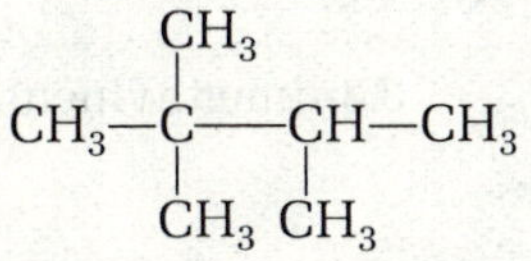
$$CH_3-\underset{\underset{CH_3}{|}}{\overset{\overset{CH_3}{|}}{C}}-\underset{\underset{CH_3}{|}}{CH}-CH_3$$

f)
$$CH_3-\underset{\underset{Br}{|}}{CH}-CH_3$$

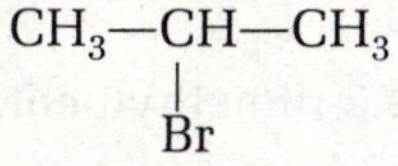

g) 1,1-dichlorocyclopropane: The root *prop* indicates three carbons; the prefix *cyclo* designates that they form a ring:

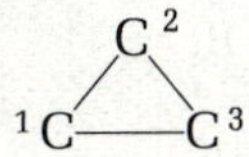

The substituents are placed:

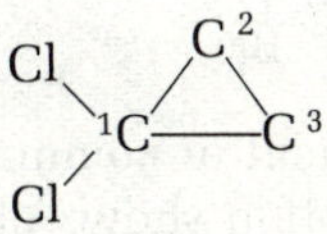

And the hydrogens are filled in:

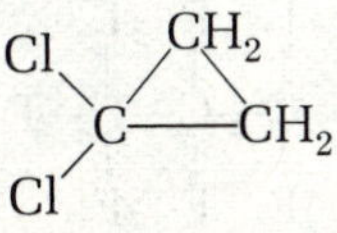

h)

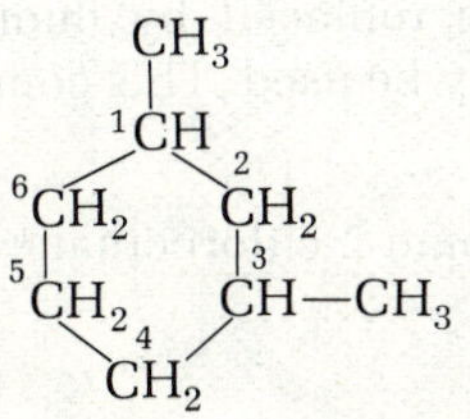

or

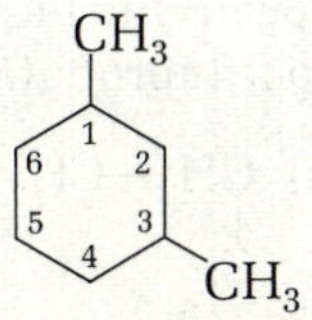

i) 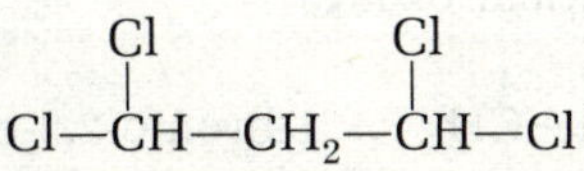
$$Cl-\underset{\underset{Cl}{|}}{CH}-CH_2-\underset{\underset{Cl}{|}}{CH}-Cl \quad \text{or} \quad CHCl_2CH_2CHCl_2$$

j)
$$CH_3-\underset{\underset{CH_3}{|}}{CH}-CH_2-CH_2-\underset{\underset{CH_3}{|}}{CH}-CH_2-CH_2-CH_3$$

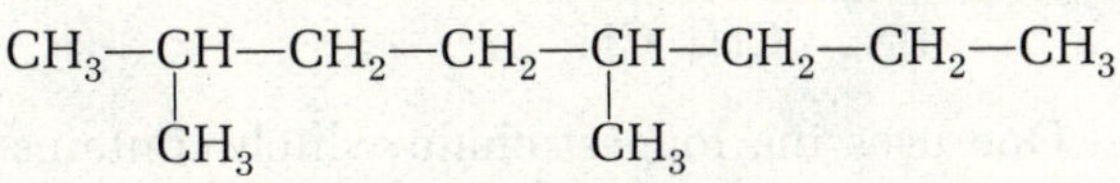

2-2. For guidance, see section 2.3.

a) $CH_3CH_2CH_2CH_2CH_3$ pentane

b) $\overset{1}{C}H_3\overset{2}{C}H\overset{3}{C}H_2\overset{4}{C}H_3$ 2-methylbutane
$\quad\quad|$
$\quad\quad CH_3$

c) $CH_3CH_2\overset{\displaystyle CH_3}{\underset{\displaystyle CH_3}{\overset{|}{\underset{|}{C}}}}CH_2CH_3$ 3,3-dimethylpentane

d) $CH_3CH_2CH_2\overset{\displaystyle CH_3}{\underset{\displaystyle CH_3}{\overset{|}{\underset{|}{C}}}}CH_3$ 2,2-dimethylpentane

e) $CH_3CH_2\overset{\displaystyle }{\underset{\displaystyle Br}{\overset{}{\underset{|}{C}}}}HCH_3$ 2-bromobutane

f) $CH_3\overset{\displaystyle Cl}{\underset{\displaystyle Cl}{\overset{|}{\underset{|}{C}}}}-\overset{\displaystyle Br}{\underset{\displaystyle Br}{\overset{|}{\underset{|}{C}}}}-Br$ 1,1,1-tribromo-2,2-dichloropropane

The placement of commas and hyphens is important; the answer to this question shows clearly how these punctuation marks are to be used.

g) $F-\overset{\displaystyle F}{\underset{\displaystyle F}{\overset{|}{\underset{|}{C}}}}-\overset{\displaystyle F}{\underset{\displaystyle F}{\overset{|}{\underset{|}{C}}}}-\overset{\displaystyle F}{\underset{\displaystyle F}{\overset{|}{\underset{|}{C}}}}-\overset{\displaystyle F}{\underset{\displaystyle F}{\overset{|}{\underset{|}{C}}}}-F$ 1,1,1,2,2,3,3,4,4,4-decafluorobutane

When all hydrogens are replaced by fluorines (or some other halogen) the term *per* may be used. This compound is also known as perfluorobutane.

h) $\overset{\displaystyle }{\underset{\displaystyle Cl}{\overset{}{\underset{|}{C}}}}H_2-\overset{\displaystyle }{\underset{\displaystyle Br}{\overset{}{\underset{|}{C}}}}H_2$ 1-bromo-2-chloroethane

(Number in alphabetical order.)

i) $\overset{1}{\underset{\displaystyle Br}{\overset{}{\underset{|}{C}}}}H_2-\overset{2}{\underset{\displaystyle CH_3}{\overset{}{\underset{|}{C}}}}H-\overset{3}{\underset{\displaystyle CH_3}{\overset{}{\underset{|}{C}}}}H-\overset{4}{C}H_3$ 1-bromo-2,3-dimethylbutane

One uses the longest chain which contains the most distinctive substituent (in this case, Br).

j)

$$CH_2$$

$$CH_2 \qquad CH_2 \qquad \text{cyclopentane}$$

$$CH_2{-}CH_2$$

2-3.

	Common	IUPAC
a)	methyl iodide	iodomethane
b)	ethyl chloride	chloroethane
c)	methylene chloride	dichloromethane
	(CH_2 = methylene)	
d)	bromoform	tribromomethane
e)	n-propyl chloride	1-chloropropane
f)	isopropyl bromide	2-bromopropane
g)	chloroform	trichloromethane
h)	hexyl bromide	bromohexane

(the position of the bromine is not specified by the formula)

i)	carbon tetrabromide	tetrabromomethane
j)	cyclobutyl chloride	chlorocyclobutane

NOTE: All the IUPAC names end in *ane*.

2-4. Write the formulas in an expanded form and then compare them. If this fails, name each by the IUPAC system and see which ones come out with the same names. For A: $a = b = f$; $c = e = h$; d is unique. For B: $c = e = g$; $b = f$; a, d, and h are unique.

2-5. a) 4-methylpentane

$$\overset{1}{C}H_3{-}\overset{2}{C}H_2{-}\overset{3}{C}H_2{-}\overset{4}{C}H{-}\overset{5}{C}H_3$$
$$|$$
$$CH_3$$

The chain should be numbered from the other end, to give the methyl substituent the lowest possible number:

$$\overset{5}{C}H_3{-}\overset{4}{C}H_2{-}\overset{3}{C}H_2{-}\overset{2}{C}H{-}\overset{1}{C}H_3 \qquad \text{2-methylpentane}$$
$$|$$
$$CH_3$$

b) 2-ethylbutane

$$\overset{1}{C}H_3{-}\overset{2}{C}H{-}\overset{3}{C}H_2{-}\overset{4}{C}H_3$$
$$|$$
$$CH_2$$
$$|$$
$$CH_3$$

The longest chain was not selected. The correct numbering is

$$CH_3{-}\overset{3}{C}H{-}\overset{4}{C}H_2{-}\overset{5}{C}H_3 \qquad \text{3-methylpentane}$$
$$|$$
$$\overset{2}{C}H_2$$
$$|$$
$$\overset{1}{C}H_3$$

13

c) Numbering started at the wrong end. The name should be 1,2-dichloropropane.

d) The ring was numbered the wrong way around to give the lowest substituent numbers. The correct name is 1,2-dimethylcyclobutane.

e) 3-methyl-3-ethylpentane

f) 2,3-dimethylhexane

g) The longest chain was not selected. The correct name is 2-methylpentane.

h) 1-bromo-2-methylpropane (use the lower number)

i) 1,1,3-trimethylcyclopentane

j) 1,1,2-trichloroethane

2-6. Approach each problem systematically, starting with the longest possible carbon chain and shortening it one carbon at a time until no further isomers are possible. To conserve space, the formulas below are written in condensed form, but the student should use expanded formulas.

a) $CH_3(CH_2)_2CH_3$ butane
   $(CH_3)_3CH$ 2-methylpropane

b) $CH_3CH_2CH_2CH_2Br$ 1-bromobutane
   $CH_3CHBrCH_2CH_3$ 2-bromobutane
   $(CH_3)_2CHCH_2Br$ 1-bromo-2-methylpropane
   $(CH_3)_3CBr$ 2-bromo-2-methylpropane

c) $CH_3(CH_2)_4CH_3$ hexane
   $CH_3CH(CH_3)CH_2CH_2CH_3$ 2-methylpentane
   $CH_3CH_2CH(CH_3)CH_2CH_3$ 3-methylpentane
   $CH_3CH(CH_3)CH(CH_3)CH_3$ 2,3-dimethylbutane
   $CH_3C(CH_3)_2CH_2CH_3$ 2,2-dimethylbutane

d) $CH_3CH_2CHBr_2$ 1,1-dibromopropane
   $CH_3CHBrCH_2Br$ 1,2-dibromopropane
   $CH_2BrCH_2CH_2Br$ 1,3-dibromopropane
   $CH_3CBr_2CH_3$ 2,2-dibromopropane

e) $CH_3CH_2CH_2CHCl_2$ 1,1-dichlorobutane
   $CH_3CH_2CHClCH_2Cl$ 1,2-dichlorobutane
   $CH_3CHClCH_2CH_2Cl$ 1,3-dichlorobutane
   $CH_2ClCH_2CH_2CH_2Cl$ 1,4-dichlorobutane
   $CH_3CCl_2CH_2CH_3$ 2,2-dichlorobutane
   $CH_3CHClCHClCH_3$ 2,3-dichlorobutane
   $CH_3CH(CH_3)CHCl_2$ 1,1-dichloro-2-methylpropane
   $CH_3CCl(CH_3)CH_2Cl$ 1,2-dichloro-2-methylpropane
   $CH_2ClCH(CH_3)CH_2Cl$ 1,3-dichloro-2-methylpropane

f) $CH_3CH_2CHBrCl$ 1-bromo-1-chloropropane
$CH_3CHClCH_2Br$ 1-bromo-2-chloropropane
$CH_2ClCH_2CH_2Br$ 1-bromo-3-chloropropane
$CH_3CHBrCH_2Cl$ 2-bromo-1-chloropropane
$CH_3CBrClCH_3$ 2-bromo-2-chloropropane

2-7. Divide the molecular weight by 12; this will tell how many carbon atoms are present.

a) $44/12 \cong 3$; ∴ the number of hydrogens is $44 - 3(12) = 8$, and the formula is C_3H_8, or propane. The monobromides are

$$CH_3CH_2CH_2Br \quad \text{and} \quad CH_3CHBrCH_3$$
1-bromopropane   2-bromopropane

b) $58/12 \cong 4$; $58 - 4(12) = 10$; ∴ C_4H_{10}. Either of two alkanes fits the substitution requirement of 2 monochloro derivatives.

$$CH_3CH_2CH_2CH_3 \longrightarrow CH_3CH_2CH_2CH_2Cl + CH_3\underset{|}{\overset{}{C}}HCH_2CH_3$$
                         Cl

1-chlorobutane   2-chlorobutane

$$CH_3\underset{CH_3}{\overset{}{C}}HCH_3 \longrightarrow CH_3\underset{CH_3}{\overset{}{C}}HCH_2Cl + CH_3-\underset{CH_3}{\overset{Cl}{C}}-CH_3$$

1-chloro-2-methylpropane   2-chloro-2-methylpropane

c) $42/12 \cong 3$; $42 - 3(12) = 6$; ∴ C_3H_6. To be an alkane, this must be monocyclic, since there are two less hydrogens than required by the formula C_nH_{2n+2}. Furthermore, to have only one monosubstitution product, all six hydrogens must be equivalent. Cyclopropane fits these requirements.

$$\underset{CH_2-CH_2}{\overset{CH_2}{\triangle}} \longrightarrow \underset{CH_2-CH_2}{\overset{\overset{Cl}{|}}{\underset{}{CH}}}$$

chlorocyclopropane

d) $72/12 = 6$; this is too much, for it does not allow any mass for the hydrogens. Therefore try 5 carbons. $72 - 5(12) = 12$; ∴ C_5H_{12}. All 12 hydrogens must be equivalent, for there is only one monochloro compound. The structures are

$$CH_3-\underset{CH_3}{\overset{CH_3}{C}}-CH_3 \longrightarrow CH_3-\underset{CH_3}{\overset{CH_3}{C}}-CH_2Cl$$

1-chloro-2,2-dimethylpropane

e) C_5H_{12} again, but four different "kinds" of hydrogens are needed. The only isomer which fits is 2-methylbutane.

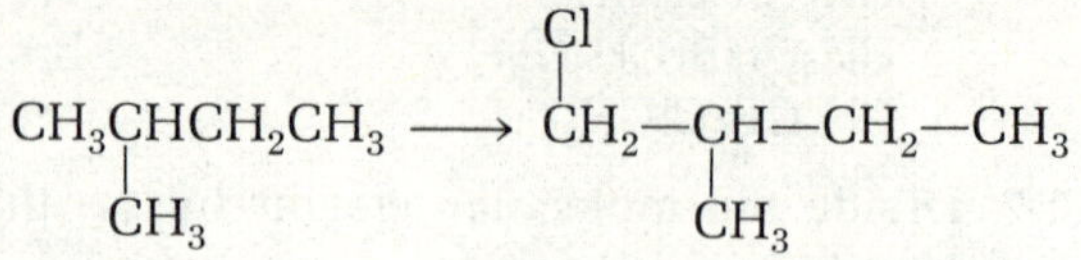

$$CH_3CHCH_2CH_3 \longrightarrow \overset{\overset{\textstyle Cl}{|}}{CH_2}-CH-CH_2-CH_3$$
$$\overset{|}{CH_3} \qquad\qquad\qquad \overset{|}{CH_3}$$

1-chloro-2-methylbutane

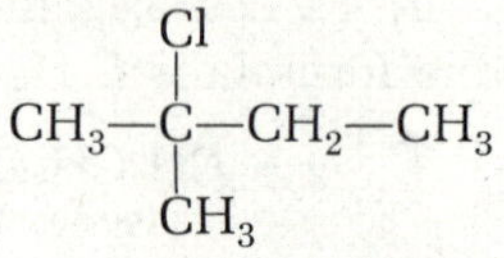

$$\overset{\overset{\textstyle Cl}{|}}{CH_3-C-CH_2-CH_3}$$
$$\overset{|}{CH_3}$$

2-chloro-2-methylbutane

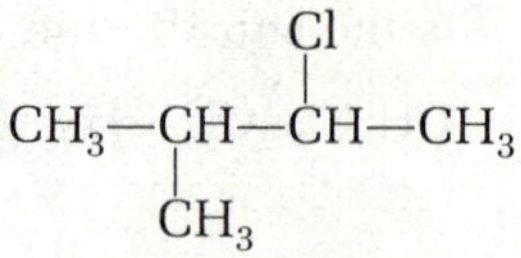

$$CH_3-CH-\overset{\overset{\textstyle Cl}{|}}{CH}-CH_3$$
$$\overset{|}{CH_3}$$

2-chloro-3-methylbutane

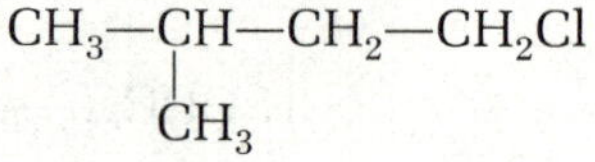

$$CH_3-CH-CH_2-CH_2Cl$$
$$\overset{|}{CH_3}$$

1-chloro-3-methylbutane

f) C_5H_{10}; there must be one ring if this is to be an alkane, and there must be only two different kinds of hydrogens. Start systematically with cyclopentane, methylcyclobutane, ethyl- or dimethylcyclopropanes. The only one which fits the substitution requirement is 1,1-dimethylcyclopropane.

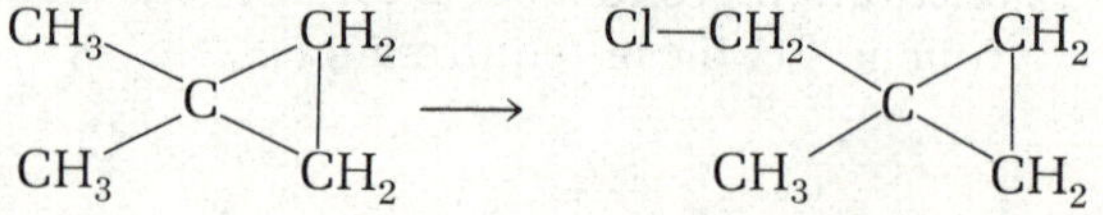

1-chloromethyl-1-methylcyclopropane

or

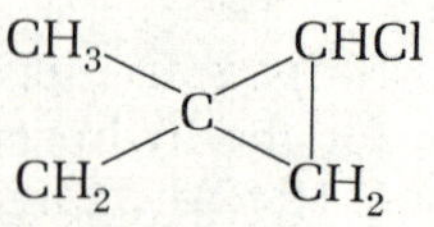

1-chloro-2,2-dimethylcyclopropane

2-8. For monochlorination an excess of hydrocarbon present at all times is preferred; therefore procedures *b* and *c*.

2-9. The order of decreasing stability is $3 > 4 > 2 > 1$. Staggered conformations are generally more stable than eclipsed, making 3 and 4 more stable than 1 and 2. Within these pairs, there are fewer methyl-methyl interactions in 3 than in 4, and in 2 than in 1.

2-10. Compare first the staggered conformers.

H

H / H

H / H

H

ethane

CH₃

H / H

H / H

H

propane

Propane should be slightly higher in energy (less stable) than ethane, due to the difference between two CH_3—H versus H—H staggered interactions. This difference should be small, since the groups are rather far apart.

Now examine the eclipsed conformers.

H

H H

H

H H

ethane

CH₃

H H

H

H H

propane

Here propane has a rather severe CH_3—H eclipsing interaction and will be very much less stable than ethane.

One concludes that the energies look like this:

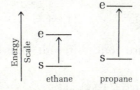

The energy difference between the two forms will be appreciably larger for propane than for ethane. Thus, as carbon chains increase in length, the preferred conformations will be extended and staggered, with rotation about the single bonds becoming gradually more difficult.

2-11. The less chain branching and the higher the molecular weight, the higher the boiling point. On these grounds, the order should be, from lowest to highest, *e, d, c, a, b*. The actual boiling points are as follows: 2-methylpentane (60°), hexane (69°), 3,3-dimethylpentane (86°), 2-methylhexane (90°), heptane (98.4°).

2-12. Note in equation 2-3 that methyl radicals are produced as reactive fragments during the chlorination of methane. If two such radicals were to combine, one would obtain ethane.

$$CH_3\cdot + CH_3\cdot \longrightarrow CH_3\!-\!CH_3 \quad \text{(ethane)}$$

This observation, then, tends to support the mechanism. The amount of ethane produced should be small, since its formation requires collision between two fragments which are present in low concentration; it is much more likely that a methyl radical will encounter a methane or chlorine molecule than another methyl radical. When the process does occur, the free radical chain is terminated.

2.13 *a)* $C_5H_{12} + 8\,O_2 \longrightarrow 5\,CO_2 + 6\,H_2O$

b) $2\,C_5H_{10} + 15\,O_2 \longrightarrow 10\,CO_2 + 10\,H_2O$

c) See equation 2-13.

d)

There are four different "kinds" of hydrogen which can be replaced.

2-14. $Cl\!-\!Cl + CH_3\!-\!H \longrightarrow CH_3\!-\!Cl + H\!-\!Cl$

Energy must be expended to break a Cl—Cl and a C—H bond; $57.8 + 87.3 = 145.1$ kcal/mole. Energy is returned by making a C—Cl and an H—Cl bond; $66.5 + 102.7 = 169.2$ kcal/mole. The reaction is exothermic (heat released) by 24.1 kcal/mole. Similar calculations show bromination is also exothermic by 7.9 kcal/mole, but iodination is endothermic (heat is absorbed by the reaction) by 6.6 kcal/mole. This reaction proceeds in the reverse direction; alkyl iodides can be reduced by HI to alkanes and iodine.

2-15. $Cl\!-\!Cl + CH_2Cl\!-\!H \longrightarrow CH_2Cl\!-\!Cl + H\!-\!Cl$

Needed to break bonds: $57.8 + 87.3 = 145.1$ kcal/mole. Gained by making bonds: $66.5 + 102.7 = 169.2$ kcal/mole. $\therefore$ reaction is exothermic by 24.1 kcal/mole. The same is true for each step in the chlorination. Heat given off at one stage could raise the reactant's temperature and accelerate the reaction, ultimately to the explosive level, if the heat produced were not removed by a heat exchanger.

2-16. $Cl\cdot + CH_4 \longrightarrow CH_3Cl + H\cdot$

Break C—H; requires 87.3 kcal/mole. Make C—Cl; gain back only 66.5 kcal/mole. ∴ this step is *endothermic* by 20.8 kcal/mole.

$$H\cdot + Cl—Cl \longrightarrow H—Cl + Cl\cdot$$

Break Cl—Cl; requires 57.8 kcal/mole. Make H—Cl; gain back 102.7 kcal/mole. ∴ this step is exothermic by 44.9 kcal/mole.

The net result is still exothermic by 24.1 kcal/mole, but the first step is strongly endothermic (requires an energy input) and is therefore a poor step in a chain reaction. This should be contrasted with the correct mechanism where both steps are exothermic:

$$Cl\cdot + CH_4 \longrightarrow H—Cl + CH_3\cdot \quad 15.4 \text{ kcal/mole} \quad \text{(exothermic)}$$

$$CH_3\cdot + Cl_2 \longrightarrow CH_3Cl + Cl\cdot \quad 8.7 \text{ kcal/mole} \quad \text{(exothermic)}$$

2-17. $CH_4 + 2 O_2 \longrightarrow CO_2 + 2 H_2O$

Break: 4 C—H bonds = 4(87.3) = 349.2
2 O=O bonds = 2(119.1) = 238.2
                 587.4 kcal/mole

Make: 2 C=O bonds = 2(192.0) = 384.0
4 H—O bonds = 4(110.6) = 442.4
                 826.4 kcal/mole

This reaction is exothermic by $826.4 - 587.4 = 239.0$ kcal/mole.

$$CH_4 + 4 Cl_2 \longrightarrow CCl_4 + 4 HCl$$

Break: 4 C—H bonds = 4(87.3) = 349.2
4 Cl—Cl bonds = 4(57.8) = 231.2
                 580.4 kcal/mole

Make: 4 C—Cl bonds = 4(66.5) = 266.0
4 H—Cl bonds = 4(102.7) = 410.8
                 676.8 kcal/mole

This reaction is exothermic but by only $676.8 - 580.4 = 96.4$ kcal/mole. The former reaction is therefore preferred.

2-18. Two alkyl groups unite at the point where the carbon-halogen bond is located in the starting halide.

   a) $CH_3CH_2CH_2—CH_2CH_2CH_3$ hexane

   b) $CH_3CH_2CH_2CH_2CH_2—CH_2CH_2CH_2CH_2CH_3$ decane

c)

dicyclopentane

d) CH_3-CH_3 .... ethane

$CH_3-CH_2CH_3$ .... propane

$CH_3CH_2-CH_2CH_3$ .... butane

A mixture which contains all three combinations of alkyl groups will be produced.

2-19. $CH_3CH_2CH_2 \{ CH_2 \{ CH_2 \{ CH_3$

$\qquad\qquad\quad a \quad\ b \quad\ c$

The two parts of the molecule may be pieced together at a, b, or c.

If at a:

$$2\,CH_3CH_2CH_2Br + 2\,Na \longrightarrow CH_3(CH_2)_4CH_3 + 2\,NaBr$$

If at b, a mixture of ethyl and butyl bromides is required:

$$CH_3CH_2CH_2CH_2Br + CH_3CH_2Br + 2\,Na \longrightarrow$$
$$CH_3(CH_2)_4CH_3 + 2\,NaBr$$

Clearly, butane and octane will also be formed, from the reaction of each halide with its own kind. Similarly at c, mixtures will be obtained (ethane, hexane, and decane).

Method a, which gives a unique product, is preferred.

2-20. This type of problem is best worked backward. Consider the Wurtz product first.

$$CH_3-\underset{\underset{CH_3}{|}}{CH}-CH_2 \{ CH_2-\underset{\underset{CH_3}{|}}{CH}-CH_3$$
$$\qquad\qquad\qquad\qquad a$$

If the two halves were joined at a, the required bromide should be 1-bromo-2-methylpropane (isobutyl bromide),

$$CH_3-\underset{\underset{CH_3}{|}}{CH}-CH_2- \{ Br$$
$$\qquad\qquad\qquad\quad a$$

and the equation would be

$$2\ CH_3CHCH_2Br + 2\ Na \longrightarrow CH_3CHCH_2CH_2CHCH_3 + 2\ NaBr$$
$$|\qquad\qquad\qquad |\qquad\qquad |$$
$$CH_3\qquad\qquad\qquad CH_3\qquad\quad CH_3$$

The Grignard reagent of this bromide would indeed give iso-butane on hydrolysis (see, for comparison, eq 2-26).

$$CH_3CH{-}CH_2Br + Mg \longrightarrow CH_3CHCH_2MgBr$$
$$|\qquad\qquad\qquad\qquad\quad |$$
$$CH_3\qquad\qquad\qquad\qquad CH_3$$

isobutylmagnesium bromide

$$CH_3CHCH_2MgBr + H{-}OH \longrightarrow CH_3CHCH_3 + Mg(OH)Br$$
$$|\qquad\qquad\qquad\qquad\qquad\qquad |$$
$$CH_3\qquad\qquad\qquad\qquad\qquad CH_3$$

2-21. *a)* The methyl group is most stable in an equatorial position:

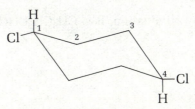

b) The *trans* arrangement (see sec 3.2) allows both chlorines to occupy equatorial positions (the remaining hydrogens are omitted, for clarity):

Note that one chlorine (at C-1) occupies the lower of the two possible bond positions, whereas the other chlorine (at C-4) uses the upper of the two bonds. Thus the chlorines are, in a sense, on opposite sides of the mean ring plane, and are *trans*.

c) The *cis* arrangement is preferred here. Both chlorines are equatorial and occupy the "upper" bonds at each carbon.

2-22. The four possible structures are

$CH_3CH_2CHCl_2$

1,1-dichloro-
propane

CH_3CH-CH_2 with Cl on both carbons

1,2-dichloro-
propane

$CH_2CH_2CH_2$ with Cl on both end carbons

1,3-dichloro-
propane

$CH_3-\underset{Cl}{\overset{Cl}{C}}-CH_3$.

2,2-dichloro-
propane

Only the last of these has all hydrogens equivalent and can give only *one* trichloro compound. This must therefore be C:

$$CH_3CCl_2CH_3 \xrightarrow{Cl_2} CH_3CCl_2CH_2Cl$$

1,3-Dichloropropane has only two different "kinds" of hydrogen. It must be D:

$$\underset{Cl}{CH_2}\underset{}{CH_2}\underset{Cl}{CH_2} \xrightarrow{Cl_2} \underset{Cl}{CH_2}-\underset{Cl}{CH}-\underset{Cl}{CH_2} \quad \text{or} \quad \underset{Cl}{CH_2}CH_2CHCl_2$$

Next, A must be capable of giving 1,2,2-trichloropropane (the product from C). This is not possible for the 1,1-isomer, since it already has two chlorines on carbon 1. Therefore A must be

$$CH_3\underset{Cl}{CH}\underset{Cl}{CH_2}$$

since it can give the 1,2,2-trichloro product (as well as 1,1,2- and 1,2,3-).

By elimination, B is $CH_3CH_2CHCl_2$.

CHAPTER THREE

UNSATURATED HYDROCARBONS: ALKENES

3-1. In each case, start with the longest possible carbon chain and determine all possible positions for the double bond. Then shorten the chain by one carbon and repeat, etc.

a) $CH_2{=}CHCH_2CH_3$ 1-butene

    $CH_3CH{=}CHCH_3$ 2-butene (*cis* and *trans*)

$CH_2{=}\underset{\underset{\displaystyle CH_3}{|}}{C}{-}CH_3$ 2-methylpropene

b) $CH_2{=}CHCH_2CH_2CH_3$ 1-pentene

$CH_3CH{=}CHCH_2CH_3$ 2-pentene (*cis* and *trans*)

$CH_2{=}\underset{\underset{\displaystyle CH_3}{|}}{C}{-}CH_2CH_3$ 2-methyl-1-butene

$CH_2{=}CH\underset{\underset{\displaystyle CH_3}{|}}{C}HCH_3$ 3-methyl-1-butene

$CH_3\underset{\underset{\displaystyle CH_3}{|}}{C}{=}CHCH_3$ 2-methyl-2-butene

c) $CH_2{=}CHCH_2CH_2CH_2CH_3$ 1-hexene

$CH_3CH{=}CHCH_2CH_2CH_3$ 2-hexene (*cis* and *trans*)

$CH_3CH_2CH{=}CHCH_2CH_3$ 3-hexene (*cis* and *trans*)

$CH_2{=}\underset{\underset{\displaystyle CH_3}{|}}{C}CH_2CH_2CH_3$ 2-methyl-1-pentene

$CH_2{=}CH\underset{\underset{\displaystyle CH_3}{|}}{C}HCH_2CH_3$ 3-methyl-1-pentene

$CH_2{=}CHCH_2\underset{\underset{\displaystyle CH_3}{|}}{C}HCH_3$ 4-methyl-1-pentene

$CH_3\underset{\underset{\displaystyle CH_3}{|}}{C}{=}CHCH_2CH_3$ 2-methyl-2-pentene

$CH_3CH{=}\underset{\underset{\displaystyle CH_3}{|}}{C}CH_2CH_3$ 3-methyl-2-pentene (*cis* and *trans*)

$CH_3CH{=}CH\underset{\underset{\displaystyle CH_3}{|}}{C}HCH_3$ 4-methyl-2-pentene (*cis* and *trans*)

$CH_2{=}\underset{\underset{\displaystyle CH_3}{|}}{C}{-}\!\!\underset{\underset{\displaystyle CH_3}{|}}{C}HCH_3$ 2,3-dimethyl-1-butene

$CH_2{=}CH{-}C(CH_3)_2CH_3$ 3,3-dimethyl-1-butene

$CH_3{-}\underset{\underset{\displaystyle CH_3}{|}}{C}{=\!\!=}\underset{\underset{\displaystyle CH_3}{|}}{C}{-}CH_3$ 2,3-dimethyl-2-butene

d) The molecular formula tells us that if this compound has only one double bond it must also have one ring. Start systematically with the largest ring and work down.

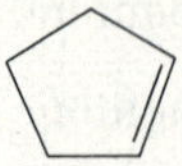

cyclopentene

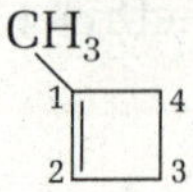

1-methylcyclobutene

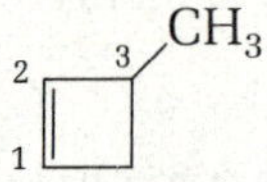

3-methylcyclobutene (one always numbers cycloalkenes "through" the double bond)

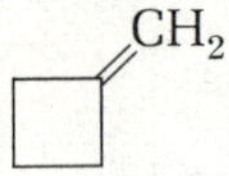

methylenecyclobutane

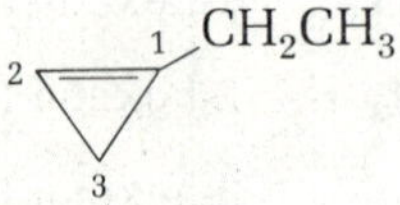

1-ethylcyclopropene

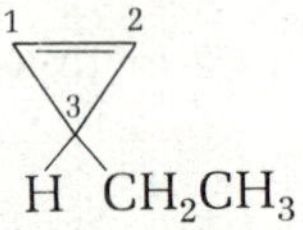

3-ethylcyclopropene

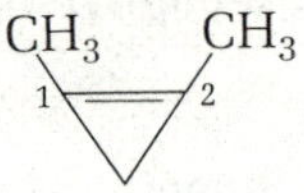

1,2-dimethylcyclopropene

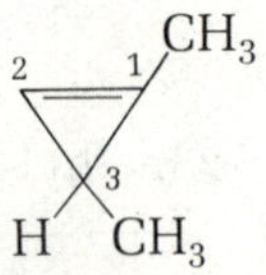

1,3-dimethylcyclopropene

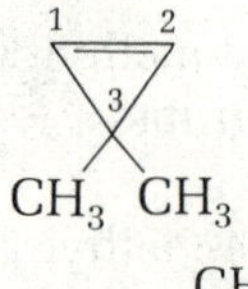

3,3-dimethylcyclopropene

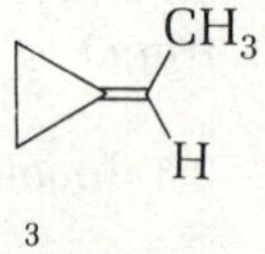

ethylidenecyclopropane

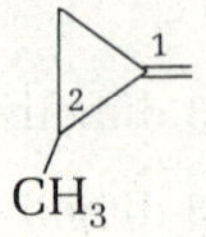

2-methylmethylenecyclopropane

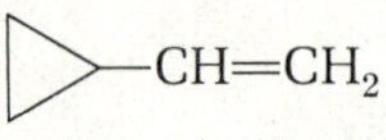

cyclopropylethene

24

3-2. *a)* 2-pentene

b) 2-methyl-2-butene (number the chain from the end with the methyl substituent).

c) 1-bromo-4-methyl-1-butene (the double bond gets the lowest possible number).

d) cyclopentene

e) *cis*-2-hexene

f) *trans*-2-hexene

g) 4-methylcyclohexene (Number "through" the double bond in such a way as to give the methyl substituent the lowest number. The doubly-bound carbons *must* be numbered 1 and 2.)

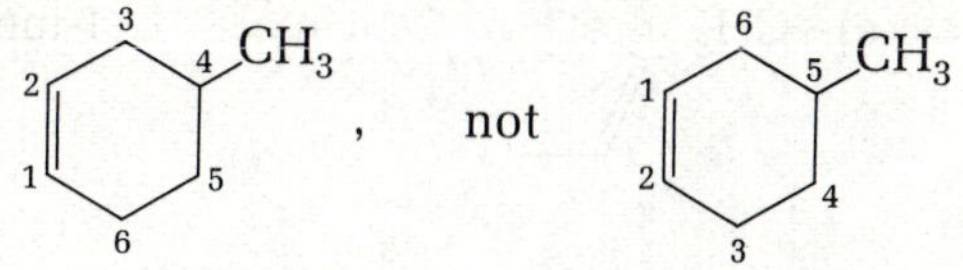

, not

3-3. In general, write out the longest carbon chain and number it, then locate the double bond, place the substituents, and finally, write in the correct number of hydrogens on each carbon atom.

a) $CH_3CH_2CH=CHCH_2CH_2CH_2CH_3$

b)
$$\overset{1}{C}H_3\overset{2}{C}H=\overset{3}{C}H-\overset{4}{C}H\overset{5}{C}H_3$$
with CH_3 below carbon 4

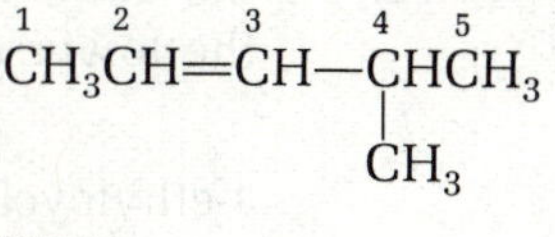

c)

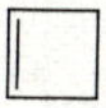

d)
$$\overset{1}{C}H_2\overset{2}{C}H=\overset{3}{C}\overset{4}{C}H_3$$
with Br below carbon 1 and Br below carbon 3

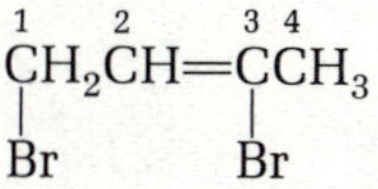

e)

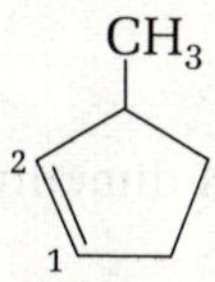

f)
$$CH_3 \quad H$$
$$\diagdown \quad \diagup$$
$$C=C$$
$$\diagup \quad \diagdown$$
$$H \qquad CH_2CH_2CH_2CH_2CH_3$$

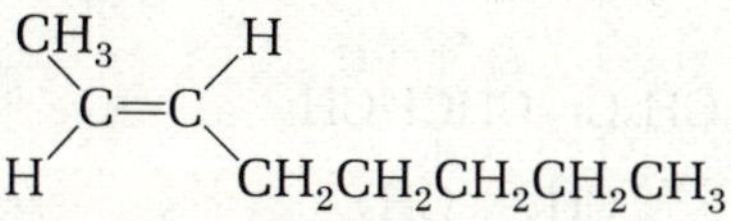

g)
$$CH_3 \quad CH_3$$
$$\diagdown \quad \diagup$$
$$C=C$$
$$\diagup \quad \diagdown$$
$$CH_3 \qquad CH_3$$

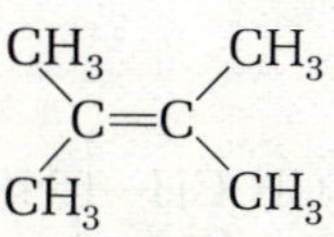

h) $CH_2=CHBr$

3-4. *a)* $\overset{1}{C}H_3\overset{2}{C}H=\overset{3}{C}H\overset{4}{C}H_3$ 2-butene; use the lower of the two numbers for the double bond.

b) $\overset{1}{C}H_3\overset{2}{C}H=\overset{3}{C}H\overset{4}{C}H_2\overset{5}{C}H_3$ 2-pentene; number as shown.

c) $\overset{1}{C}H_2=\overset{2}{C}-CH_3$
$\quad\quad\overset{3}{|}\quad\overset{4}{}$
$\quad\quad CH_2CH_3$

2-methyl-1-butene; number the longest chain.

d)
$\quad\quad Cl$
$\quad\quad \overset{3}{|}\quad\overset{2}{}\quad\overset{1}{}$
$CH_3-\overset{}{C}-CH=CH_2$
$\quad\quad\overset{4}{|}\quad\overset{5}{}$
$\quad\quad CH_2CH_3$

3-chloro-3-methyl-1-pentene

e) CH_3

1-methylcyclopentene

not

f)
$\quad\quad\quad CH_3$
$\overset{5}{}\quad\overset{4}{|}\quad\overset{3}{}\quad\overset{2}{}\quad\overset{1}{}$
$CH_3-C-CH_2CH=CH_2$
$\quad\quad |$
$\quad\quad CH_3$

4,4-dimethyl-1-pentene; the double bond gets the lower number

g) CH_3CH_2

3-ethylcyclobutene

not CH_3CH_2

h) $\overset{6}{C}H_3\overset{5}{C}H_2\overset{4}{C}=\overset{3}{C}H\overset{2}{C}H\overset{1}{C}H_3$
$\quad\quad\quad\quad |\quad\quad |$
$\quad\quad\quad\quad CH_3\quad CH_3$

2,4-dimethyl-3-hexene

not $\overset{1}{C}H_3\overset{2}{C}H_2\overset{3}{C}=\overset{4}{C}H\overset{5}{C}H\overset{6}{C}H_3$
$\quad\quad\quad\quad\quad |\quad\quad |$
$\quad\quad\quad\quad\quad CH_3\quad CH_3$

3-5. Review equation 3-5.

a) $CH_3CH=CHCH_3 \xrightarrow{Br_2} CH_3CH-CHCH_3$
$\quad\quad\quad\quad\quad\quad\quad\quad\quad\quad\quad\quad\quad | \quad\quad |$
$\quad\quad\quad\quad\quad\quad\quad\quad\quad\quad\quad\quad Br\quad Br$

2,3-dibromo-butane

b) $CH_3CH_2\underset{\underset{\displaystyle CH_3}{|}}{C}=CHCH_2CH_3 \xrightarrow{Br_2}$

$$\overset{1}{C}H_3\overset{2}{C}H_2\underset{\underset{\displaystyle CH_3}{|}}{\overset{3}{\underset{\displaystyle |}{\overset{\displaystyle Br}{C}}}}-\overset{4}{\underset{\displaystyle |}{\overset{\displaystyle Br}{C}}}H\overset{5}{C}H_2\overset{6}{C}H_3$$

3,4-dibromo-3-methylhexane

c) $CH_2=CHCl \xrightarrow{Br_2} \underset{\underset{\displaystyle Br}{|}}{\overset{2}{C}}H_2-\underset{\underset{\displaystyle Br}{|}}{\overset{1}{C}}HCl$

1,2-dibromo-1-chloroethane

d)

1,2-dibromo-1-methylcyclopentane

3-6. a) $CH_2=CHCH_2CH_3 + Cl_2 \longrightarrow \underset{\underset{\displaystyle Cl}{|}}{C}H_2-\underset{\underset{\displaystyle Cl}{|}}{C}HCH_2CH_3$ (eq 3-9)

b) $CH_2=CHCH_2CH_3 + HCl \longrightarrow CH_3\underset{\underset{\displaystyle Cl}{|}}{C}HCH_2CH_3$ (eq 3-14)

c) $CH_2=CHCH_2CH_3 + HOBr \longrightarrow$

$\underset{\underset{\displaystyle Br}{|}}{C}H_2-\underset{\underset{\displaystyle OH}{|}}{C}HCH_2CH_3$ (eq 3-17)

d) $CH_2=CHCH_2CH_3 + H-OSO_3H \longrightarrow$

$CH_3\underset{\underset{\displaystyle OSO_3H}{|}}{C}HCH_2CH_3$ (eq 3-15)

e) $CH_2=CHCH_2CH_3 + O_3 \longrightarrow CH_2=O + O=CHCH_2CH_3$

(eq 3-26)

f) $CH_2=CHCH_2CH_3 \xrightarrow{KMnO_4} \underset{\underset{\displaystyle OH}{|}}{C}H_2-\underset{\underset{\displaystyle OH}{|}}{C}HCH_2CH_3$ (eq 3-25)

g) $CH_2=CHCH_2CH_3 \xrightarrow[K^+Cl^-]{Br_2} \underset{\underset{\displaystyle Br}{|}}{C}H_2-\underset{\underset{\displaystyle Br}{|}}{C}HCH_2CH_3$ and

$\underset{\underset{\displaystyle Br}{|}}{C}H_2-\underset{\underset{\displaystyle Cl}{|}}{C}HCH_2CH_3$ (eq 3-6)

3-7. See each of the following sections:

a) 3.1   e) 3.4a

b) 3.4   f) 3.4e

c) 3.4   g) 3.2

d) 3.4c   h) 3.4d

3-8. a) $CH_2{=}CHCH_3 + H_2 \xrightarrow[\text{catalyst}]{\text{Pt}} CH_3CH_2CH_3$ (eq 3-10)

b) $CH_2{=}CHCH_3 + HBr \longrightarrow CH_3\underset{\underset{\displaystyle Br}{|}}{C}HCH_3$ (eq 3-14)

c) $CH_2{=}CHCH_3 + Cl_2 \longrightarrow \underset{\underset{\displaystyle Cl}{|}}{C}H_2{-}\underset{\underset{\displaystyle Cl}{|}}{C}HCH_3$ (eq 3-9)

d) $CH_2{=}CHCH_3 + HOCl \longrightarrow \underset{\underset{\displaystyle Cl}{|}}{C}H_2{-}\underset{\underset{\displaystyle OH}{|}}{C}HCH_3$ (eq 3-17)

e) $CH_2{=}CH{-}CH_3 \xrightarrow[\text{(eq 3-14)}]{\text{HBr}} CH_3\underset{\underset{\displaystyle Br}{|}}{C}HCH_3 \xrightarrow[\text{(eq 2-22)}]{\text{Na}}$

$CH_3\underset{\underset{\displaystyle CH_3}{|}}{C}H{-}\underset{\underset{\displaystyle CH_3}{|}}{C}HCH_3$

f) $CH_2{=}CHCH_3 + HOSO_3H \longrightarrow CH_3\underset{\underset{\displaystyle OSO_3H}{|}}{C}HCH_3$ (eq 3-15)

3-9. The other product would be 1-bromo-2-chloropropane, since reaction is initiated by Br^+ and follows Markownikoff's rule.

$CH_2{=}CHCH_3 \xrightarrow{Br^+} \underset{\underset{\displaystyle Br}{|}}{C}H_2\overset{+}{C}HCH_3 \xrightarrow{Cl^-} \underset{\underset{\displaystyle Br}{|}}{C}H_2\underset{\underset{\displaystyle Cl}{|}}{C}HCH_3$

3-10. Reaction is initiated by the electrophile Br^+. The product of

the first step may react with any nucleophile, including methanol. The two products will be

3-11. *a)* $CH_3CH_2CH_2Br \xrightarrow[\substack{KOH \\ eq\ 3\text{-}31}]{alc.} CH_3CH{=}CH_2 \xrightarrow[eq\ 3\text{-}5]{Br_2} \underset{\substack{| \\ Br}}{CH_2}{-}\underset{\substack{| \\ Br}}{CHCH_3}$

b) $\underset{\substack{| \\ OH}}{CH_3CHCH_3} \xrightarrow[eq\ 3\text{-}29]{H_2SO_4} CH_2{=}CHCH_3 \xrightarrow[eq\ 3\text{-}14]{HI} \underset{\substack{| \\ I}}{CH_3CHCH_3}$

c) $CH_3CH_2CH_2CH_2Br \xrightarrow[\substack{KOH \\ eq\ 3\text{-}31}]{alc.} CH_3CH_2CH{=}CH_2 \xrightarrow[eq\ 3\text{-}14]{HCl}$

$$\underset{\substack{| \\ Cl}}{CH_3CH_2CHCH_3}$$

d) $CH_3CH_2Br \xrightarrow[\substack{KOH \\ eq\ 3\text{-}31}]{alc.} CH_2{=}CH_2 \xrightarrow[eq\ 3\text{-}17]{HOCl} \underset{\substack{| \\ Cl}}{CH_2}{-}\underset{\substack{| \\ OH}}{CH_2}$

e) $\underset{\substack{| \\ OH}}{CH_3CH_2CHCH_3} \xrightarrow[eq\ 3\text{-}30]{H^+}$

$$\left.\begin{array}{c} CH_3CH{=}CHCH_3 \\ + \\ CH_3CH_2CH{=}CH_2 \end{array}\right] \xrightarrow[\substack{Pt \\ eq\ 3\text{-}10}]{H_2} CH_3CH_2CH_2CH_3$$

f) $\xrightarrow[\substack{KOH \\ eq\ 3\text{-}31}]{alc.}$ $\xrightarrow[eq\ 3\text{-}5]{Br_2}$

3-12. First calculate the number of moles of alkene; this is equal to the number of moles of bromine required (eq 3-5). Then multiply by the molecular weight of bromine.

a) 14 g of 2-butene = $14/C_4H_8 = 14/56 = 0.25$ moles; $\therefore$ need 0.25 mole of $Br_2 = 0.25 \times 160 = 40$ g.

b) $41/C_6H_{10} = 41/82 = 0.5$ moles $\therefore$ need $0.5 \times 160 = 80$ g of bromine.

3-13. 8.0 g of bromine = $8/160 = 0.05$ moles. Therefore there must be 0.05 mole of 2-butene present. This is $0.05 \times C_4H_8 = 0.05 \times 56 = 2.8$ g of 2-butene. Therefore there must be $10.0 - 2.8 = 7.2$ g of butane or 72% of butane in the original mixture.

3-14. Review sec 3.2 before answering this question.

a) Only one structure is possible.

b)

cis-2-pentene *trans*-2-pentene

c) Only one structure.

d)

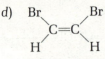

cis-1,2-dibromoethene   trans-1,2-dibromoethene

e)

Cl   CH₃   Cl   H
  \   /   \   /
   C=C   C=C
  /   \   /   \
H   H   H   CH₃

cis-1-chloropropene   trans-1-chloropropene

f) Only one structure.

g) Only one structure:

h)

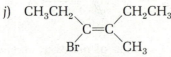

cis-1,2-dibromocyclopentane   trans-1,2-dibromocyclopentane

i) Only one structure.

j)

CH₃CH₂   CH₂CH₃   CH₃CH₂   CH₃
     \   /   \   /
      C=C   C=C
     /   \   /   \
   Br   CH₃   Br   CH₂CH₃

cis-3-bromo-4-methyl-   trans-3-bromo-4-methyl-3-hexene
3-hexene

3-15. When the ring is sufficiently large, both *cis* and *trans* isomers are possible:

cis-cyclodecene   trans-cyclodecene

If the ring is small, as in cyclohexene, only the *cis* structure is geometrically possible:

cyclohexene
(*cis*)

trans-cyclohexene;
an impossible structure

trans-cyclohexene is an impossible structure because the four carbons marked • must lie in a plane; two CH_2 groups are not enough to reach between the ends of this system.

3-16. In each case, note the substituents, and place a double bond in the starting material at the appropriate position:

   a) $CH_3CH_2CH{=}CH_2 + Br_2$ (eq 3-5)

   b) $CH_3CH{=}CHCH_3 + HOCl$ (eq 3-17)

   c) $CH_3\overset{|}{\underset{CH_3}{C}}{=}CH_2 + H_2$ (Pt catalyst)   (eq 3-10)

   d) $CH_2{=}CHCH_2CH_2CH_3 + HBr$(eq 3-14; if one selected the other possible precursor, $CH_3CH{=}CHCH_2CH_3$, one would obtain a mixture of the desired product plus $CH_3CH_2CHBrCH_2CH_3$).

   e) $CH_2{=}CHCH_3 + HOSO_3H$ (eq 3-15).

3-17. One must select a symmetric alcohol which can give only a single product:

   a) $CH_3-\overset{\overset{\displaystyle OH}{|}}{\underset{\underset{\displaystyle CH_3}{|}}{C}}-CH_3$

   b) $CH_3CH_2CH_2\overset{}{\underset{\underset{\displaystyle OH}{|}}{C}}HCH_2CH_2CH_3$

   c)

   d) $CH_3-$$-OH$

3-18. The alkene which gave the particular aldehyde or ketone ($C{=}O$ compound) can be deduced by joining the two carbons attached to O's by a $C{=}C$ double bond:

   a) $CH_3CH_2CH{=}CHCH_2CH_3$

   b) $(CH_3)_2C{=}CHCH_3$

   c) $CH_2{=}CHCH(CH_3)_2$

   d) $\underset{\displaystyle CH_2-CH}{\overset{\displaystyle CH_2-CH}{|\qquad\|}}$

3-19. Use equation 3-18 as a starting point.

This is a tertiary carbonium ion; the other alternative,

would be a secondary carbonium ion and is less favored.

3-20. In each case, the secondary carbonium ion is formed in preference to the primary carbonium ion, in accord with Markownikoff's rule:

$$CH_2{=}CHCH_3 + H^+ \longrightarrow CH_3\overset{+}{C}HCH_3$$

The isopropyl cation adds to another propene molecule:

$$CH_3\overset{+}{C}HCH_3 + CH_2{=}CHCH_3 \longrightarrow CH_3\underset{\underset{CH_3}{|}}{C}H{-}CH_2\overset{+}{C}HCH_3$$

The process is repeated:

$$CH_3\underset{\underset{CH^3}{|}}{C}HCH_2\overset{+}{\underset{\underset{CH_3}{|}}{C}}H + CH_2{=}CH{-}CH_3 \longrightarrow$$

$$CH_3\underset{\underset{CH_3}{|}}{C}HCH_2\underset{\underset{CH_3}{|}}{C}HCH_2\overset{+}{C}HCH_3$$

And once more:

$$CH_3\underset{\underset{CH_3}{|}}{C}HCH_2\underset{\underset{CH_3}{|}}{C}HCH_2\overset{+}{\underset{\underset{CH_3}{|}}{C}}H + CH_2{=}CHCH_3 \longrightarrow$$

$$CH_3\underset{\underset{CH_3}{|}}{C}HCH_2\underset{\underset{CH_3}{|}}{C}HCH_2\underset{\underset{CH_3}{|}}{C}HCH_2\overset{+}{C}HCH_3$$

In the final step, a proton is lost. This can give either the product shown in eq 3-36 or its double-bond isomer.

$$CH_3CHCH_2CHCH_2CHCH_2\overset{+}{C}HCH_3 \quad\begin{cases} \longrightarrow CH_3CHCH_2CHCH_2CHCH=CHCH_3 \\ CH_3 \quad CH_3 \quad CH_3 \\ \\ \longrightarrow CH_3CHCH_2CHCH_2CHCH_2CH=CH_2 \\ CH_3 \quad CH_3 \quad CH_3 \end{cases}$$

CH₃   CH₃   CH₃

CHAPTER FOUR

UNSATURATED HYDROCARBONS: ALKYNES, DIENES, AND POLYENES

4-1. In each case, the alkynes and then the alkadienes are given. Geometric isomers are not designated.

a) $HC\equiv C-CH_2CH_2CH_3$ — 1-pentyne

$CH_3C\equiv CCH_2CH_3$ — 2-pentyne

$HC\equiv CCHCH_3$ — 3-methyl-1-butyne

  $\quad\quad\quad\ CH_3$

$CH_2=C=CH-CH_2CH_3$ — 1,2-pentadiene

$CH_2=CH-CH=CH-CH_3$ — 1,3-pentadiene

$CH_2=CH-CH_2-CH=CH_2$ — 1,4-pentadiene

$CH_3-CH=C=CH-CH_3$ — 2,3-pentadiene

$CH_2=C=C-CH_3$ — 3-methyl-1,2-butadiene

  $\quad\quad\quad\ CH_3$

$CH_2=C-CH=CH_2$ — 2-methyl-1,3-butadiene

  $\quad\ CH_3$

b) $HC\equiv CCH_2CH_2CH_2CH_3$ — 1-hexyne

$CH_3C\equiv CCH_2CH_2CH_3$ — 2-hexyne

$CH_3CH_2C\equiv CCH_2CH_3$ — 3-hexyne

$HC\equiv CCHCH_2CH_3$ — 3-methyl-1-pentyne

  $\quad\quad\quad\ CH_3$

$HC\equiv CCH_2CHCH_3$ — 4-methyl-1-pentyne

  $\quad\quad\quad\quad\ CH_3$

$CH_3C\equiv CCHCH_3$ — 4-methyl-2-pentyne

  $\quad\quad\quad\ CH_3$

$$\underset{\underset{CH_3}{|}}{\overset{\overset{CH_3}{|}}{HC{\equiv}C}}CCH_3$$

3,3-dimethyl-1-butyne

$$CH_2{=}C{=}CHCH_2CH_2CH_3$$

1,2-hexadiene (also the 1,3-; 1,4-; 1,5-; 2,3-; 2,4-isomers)

$$\underset{\underset{CH_3}{|}}{CH_2{=}C{=}C}CH_2CH_3$$

3-methyl-1,2-pentadiene (also the 4-methyl isomer)

$$\underset{\underset{CH_3}{|}}{CH_2{=}C}{-}CH{=}CHCH_3$$

2-methyl-1,3-pentadiene (also the 3- and 4-methyl isomers)

$$\underset{\underset{CH_3}{|}}{CH_2{=}C}{-}CH_2{-}CH{=}CH_2$$

2-methyl-1,4-pentadiene (also the 3-methyl isomer)

$$\underset{\underset{CH_3}{|}}{CH_3C}{=}C{=}CHCH_3$$

2-methyl-2,3-pentadiene

$$\underset{\underset{CH_2CH_3}{|}}{CH_2{=}C}{-}CH{=}CH_2$$

2-ethyl-1,3-butadiene (NOTE: in this example the longest chain which includes both double bonds is *not* the longest carbon chain in the molecule; nevertheless it is selected for the root of the name because it does include both double bonds.)

$$\underset{\underset{CH_3}{|}}{CH_2{=}C}{-}\underset{\underset{CH_3}{|}}{C}{=}CH_2$$

2,3-dimethyl-1,3-butadiene

c) The empirical formula tells us that the molecule must contain one of the following combinations: (a) a triple bond + a double bond, (b) a triple bond + a ring, (c) three double bonds, (d) two double bonds + a ring, (e) one double bond + two rings. Taking the terms alkyne and alkadiene to imply only one triple bond or two double bonds as the sites of unsaturation, we will give answers in categories (b) and (d) only. The reader may wish to give all possible answers in the other three categories.

$$\triangleright{-}C{\equiv}CH$$ cyclopropyl acetylene (not a IUPAC name)

(The triple bond cannot be in a small ring—i.e., cyclopentyne—because of its linear geometry.)

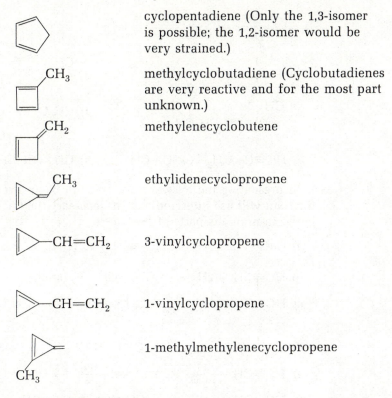

cyclopentadiene (Only the 1,3-isomer is possible; the 1,2-isomer would be very strained.)

methylcyclobutadiene (Cyclobutadienes are very reactive and for the most part unknown.)

methylenecyclobutene

ethylidenecyclopropene

3-vinylcyclopropene

1-vinylcyclopropene

1-methylmethylenecyclopropene

4-2. *a*) 3-methyl-1-butyne

   b) 2-pentyne

   c) 2-bromo-1,3-butadiene

   d) 2-methyl-1,3-cyclohexadiene; the ring is numbered through the two double bonds.

$$CH_3 \; \overset{2}{\diagdown} \quad \overset{1}{\diagup}$$

   e) 2,4-dimethyl-2,3-pentadiene (also, tetramethylallene)

   f) butenyne (also, vinylacetylene)

   g) cyclodecyne (The large ring can accommodate the linear triple bond without undue strain.)

4-3. *a*) $\overset{1}{CH_2}=\overset{2}{CH}\overset{3}{CH_2}\overset{4}{CH}=\overset{5}{CH}\overset{6}{CH_3}$ *b*) $HC\equiv C-CH-CH_2-CH_3$
$$\qquad\qquad\qquad\qquad\qquad\qquad\qquad\qquad\quad |$$
$$\qquad\qquad\qquad\qquad\qquad\qquad\qquad\qquad\; CH_3$$

c) (cyclopentyl)—CH=CH$_2$

$$f)\ CH_3 \overset{2}{\underset{3}{\diagdown}} \overset{1}{\underset{4}{\diagup}}^{5}$$
CH$_3$

d)
$$\underset{\underset{\underset{\text{CH}_2\text{CH}_2\text{CH}_2\text{CH}_2}{}}{\overset{\text{H}}{\underset{\text{CH}_2}{\text{C}}}}}{\text{H}}{=}\text{C}{=}\overset{\text{H}}{\underset{\text{CH}_2}{\text{C}}}$$

$$g)\ CH_3C{\equiv}C{-}\overset{\overset{\displaystyle CH_3}{|}}{\underset{\underset{\displaystyle CH_3}{|}}{C}}{-}{-}CHCH_3$$
CH$_3$

e) HC≡C—CH$_2$—C≡C—CH$_3$ h) CH$_2$=CHCH$_2$Br

4-4. In presenting the answers, we will not begin from CaC$_2$ each time but will use intermediates, if necessary, that have been prepared in previous parts of the question.

a) CaC$_2$ + 2H$_2$O $\longrightarrow$ HC≡CH + Ca(OH)$_2$ (eq 4-17)

b) HC≡CH + 2H$_2$ $\xrightarrow{\text{Pt}}$ CH$_3$—CH$_3$ (analogous to eq 3-10)

c) HC≡CH + H$_2$ $\xrightarrow[\substack{\text{catalyst} \\ \text{eq 4-3}}]{\text{Pd}}$ CH$_2$=CH$_2$ $\xrightarrow[\text{eq 3-11}]{\text{HI}}$ CH$_3$CH$_2$I

d) 2HC≡CH $\xrightarrow[\text{NH}_4\text{Cl}]{\text{Cu}_2\text{Cl}_2}$ HC≡C—CH=CH$_2$ (eq 4-33)

e) HC≡CH $\xrightarrow[\text{eq 4-10}]{\text{NaNH}_2}$ HC≡C$^-$Na$^+$ $\xrightarrow[\text{eq 4-11}]{\text{CH}_3\text{CH}_2\text{I}}$ HC≡CCH$_2$CH$_3$

f) HC≡CH + 2HI $\longrightarrow$ CH$_3$CHI$_2$ (eq 4-5)

g) HC≡CCH$_2$CH$_3$ (from part e) + 2HBr $\longrightarrow$ CH$_3$CBr$_2$CH$_2$CH$_3$

(eq 4-5)

h) HC≡CCH$_2$CH$_3$ $\xrightarrow[\text{eq 4-10}]{\text{NaNH}_2}$ Na$^+$ $^-$C≡CCH$_2$CH$_3$ $\xrightarrow[\text{eq 4-11}]{\text{CH}_3\text{CH}_2\text{I}}$

CH$_3$CH$_2$C≡CCH$_2$CH$_3$

i) HC≡CH + 2Br$_2$ $\longrightarrow$ CHBr$_2$CHBr$_2$ (eq 4-2)

j) CH$_3$CH$_2$C≡CCH$_2$CH$_3$ (from h) + H$_2$ $\xrightarrow[\substack{\text{catalyst} \\ \text{eq 4-3}}]{\text{Pd}}$

CH$_3$CH$_2$CH=CHCH$_2$CH$_3$
(cis isomer)

4-5. In problems of this type one should carefully examine the structures of the starting material and the final product. Seek out similarities and differences. Note the types of bonds which must be made or broken in order to go from one structure to the other. Sometimes it is profitable to work backward from the product, and forward from the starting material simultaneously, hoping to arrive at a common intermediate.

a) $CH_3CH{=}CHCH_3 \xrightarrow[\text{eq 3-2}]{Br_2}$

$$CH_3\underset{\underset{Br}{|}}{CH}{-}\underset{\underset{Br}{|}}{CH}CH_3 \xrightarrow[\text{eq 3-31}]{\underset{\text{in alcohol}}{KOH}} CH_2{=}CH{-}CH{=}CH_2$$

(Possible by-products in the second step include $CH_3C{\equiv}CCH_3$ and $CH_2{=}C{=}CHCH_3$.)

b) $CH_3\underset{\underset{OH}{|}}{CH}CH_3 \xrightarrow[\text{eq 3-29}]{\overset{H^+}{\text{heat}}} CH_2{=}CHCH_3 \xrightarrow[\text{eq 3-2}]{Br_2}$

$$\underset{\underset{Br}{|}}{CH_2}{-}\underset{\underset{Br}{|}}{CH}{-}CH_3 \xrightarrow[\text{eq 4-19}]{\underset{\text{in alcohol}}{KOH}} HC{\equiv}CCH_3$$

(A possible by-product in the last step is $CH_2{=}C{=}CH_2$.)

c) $CH_3CH_2CH_2Br \xrightarrow[\text{eq 3-31}]{\underset{\text{in alcohol}}{KOH}} CH_3CH{=}CH_2 \xrightarrow[\text{part b}]{\text{as in}}$

$$CH_3C{\equiv}CH \xrightarrow[\text{eq 4-5}]{2\,HCl} CH_3\underset{\underset{Cl}{|}}{\overset{\overset{Cl}{|}}{C}}{-}CH_3$$

d) $CH_2{=}CH{-}CH{=}CH_2 \xrightarrow[\text{eq 4-22}]{Br_2}$

$$\underset{\underset{Br}{|}}{CH_2}{-}CH{=}CH{-}\underset{\underset{Br}{|}}{CH_2} \xrightarrow[\text{eq 3-10}]{\overset{H_2}{Pt}} \underset{\underset{Br}{|}}{CH_2}CH_2CH_2\underset{\underset{Br}{|}}{CH_2}$$

e) $CH_2{=}CHCH_2CH_3 \xrightarrow[\text{eq 3-2}]{Br_2}$

$$\underset{\underset{Br}{|}}{CH_2}{-}\underset{\underset{Br}{|}}{CH}CH_2CH_3 \xrightarrow[\text{eq 4-19}]{\underset{\text{in alcohol}}{KOH}} HC{\equiv}C{-}CH_2CH_3$$

4-6. Often in this type of problem it is fruitful to work backward from the last piece of information given.

Fact *d* tells us that the structure $HC{\equiv}C{-}$ is present (eq 4-15), and fact *c* tells us that this is the only unsaturation present in the molecule:

$$HC{\equiv}C{-}\; +\; 2\,Br_2 \longrightarrow H\underset{\underset{Br}{|}}{\overset{\overset{Br}{|}}{C}}{-}\underset{\underset{Br}{|}}{\overset{\overset{Br}{|}}{C}}{-}$$

Fact *b* enables us to calculate the empirical formula:

$$\text{C} \quad 88/12 = 7.33 \qquad 7.33/7.33 = 1.0 \quad = \frac{2.0}{3.28} = \frac{3.0}{4.92}$$
$$\text{H} \quad 12/1 = 12.00 \qquad 12.00/7.33 = 1.64 \quad 3.28 4.92$$

This suggests an empirical formula C_3H_5. The molecular weight given as fact *a* tells us the correct formula is C_6H_{10} (M.W. = 82). Actually we could have "guessed" this from the molecular weight and the information that the molecule contains only carbon and hydrogen (Fact *b*). Note that the formula C_6H_{10} is consistent with the acetylenic structure (C_nH_{2n-2}).

One possible structure is $HC{\equiv}CCH_2CH_2CH_2CH_3$. Can you draw three others?

4-7. Rotation would require that the π bonds be broken; however, after only a 90° rotation the *p* orbitals would be properly aligned for rebonding (with a double bond, 180° rotation is required). One therefore expects that rotation about a triple bond would not be "free" but would be considerably more facile than rotation about a double bond. Since the bond is linear, no geometric isomerism is possible, and the phenomenon of rotation about a triple bond is difficult to study experimentally.

4-8. The mechanism is analogous to that for bromine addition (sec 4.8).

$$CH_2{=}CH{-}CH{=}CH_2 + H^+ \longrightarrow$$

$$[CH_3\overset{+}{C}H{-}CH{=}CH_2 \longleftrightarrow CH_3CH{=}CH{-}\overset{+}{C}H_2] \xrightarrow{HBr}$$

CH₃CH—CH=CH₂ + CH₃CH=CH—CH₂ + H⁺
     | |
     Br   Br

3-bromo-1-butene   1-bromo-2-butene

4-9. *a)* CH₂=C—CH=CH₂ + Br₂ ⟶ CH₂—C=CH—CH₂
              | | | |
              CH₃   Br   CH₃   Br

1,4-Addition predominates; see section 4.8.

b)

This is an example of the Diels-Alder reaction, section 4.9.

c) $HC\equiv C^-Na^+ + CH_3CHCH_2CH_3 \xrightarrow[\text{eq 4-11}]{S_N2}$
$\qquad\qquad\qquad\qquad\;\; |$
$\qquad\qquad\qquad\qquad\;\; Br$

$$HC\equiv C-CHCH_2CH_3 + Na^+Br^-$$
$$\qquad\qquad\quad |$$
$$\qquad\qquad\quad CH_3$$

d)

$\xrightarrow[\text{alcohol}]{\text{KOH in}}$

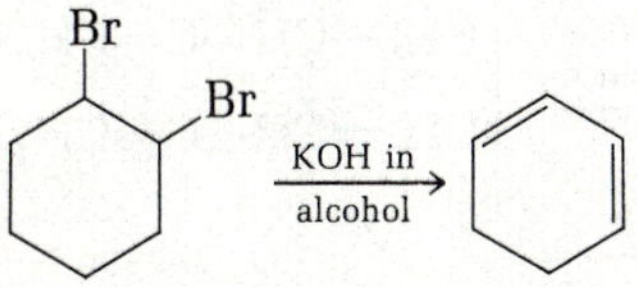

(eq 3-31)

It is not possible to put a triple bond in the six-membered ring.

e)

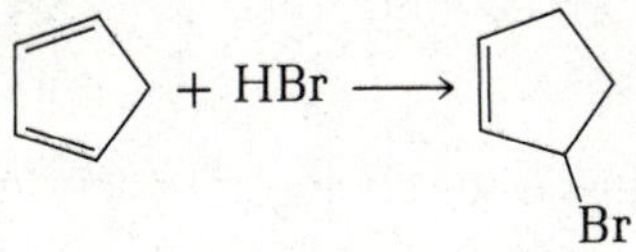

$+\ HBr \longrightarrow$

The product is the same, whether by 1,2- or 1,4-addition.

f)

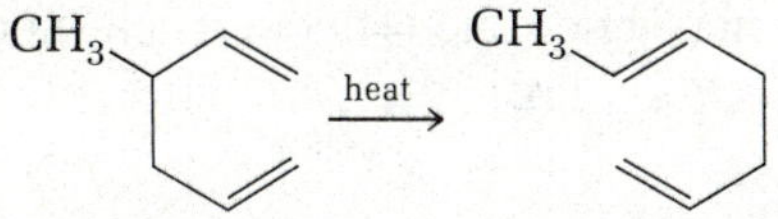

$\xrightarrow{\text{heat}}$

This is a Cope rearrangement, equation 4-28.

4-10. Cyclopentadiene can undergo a Diels-Alder reaction with itself, in which one molecule acts as diene and the other as dienophile; at higher temperatures, the reaction proceeds in the reverse direction:

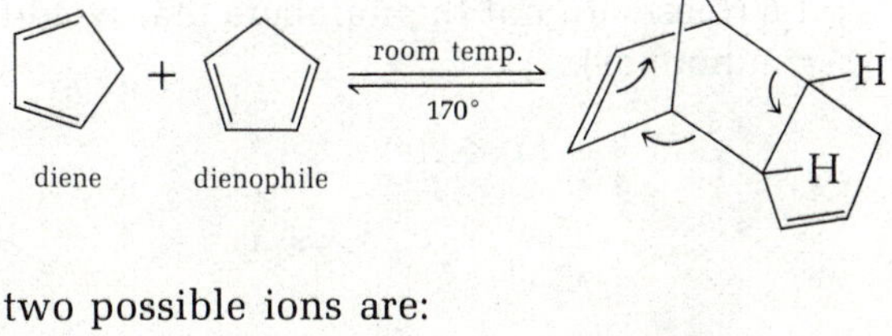

diene   dienophile $\underset{170°}{\overset{\text{room temp.}}{\rightleftharpoons}}$

4-11. The two possible ions are:

$CH_2=CH-CH=CH_2$

attack at C1 $\longrightarrow$ $\begin{bmatrix} CH_3\overset{+}{C}H-CH=CH_2 \\ \updownarrow \\ CH_3CH=CH-\overset{+}{C}H_2 \end{bmatrix}$

attack at C2 $\longrightarrow$ $\overset{+}{C}H_2-CH_2-CH=CH_2$

Attack at C1 gives an allylic, resonance-stabilized carbonium ion;

this is preferred over attack at C2 which leads to the relatively unstable primary carbonium ion.

4-12. The immediate precursor to chloroprene in eq 4-33 is an allylic chloride. If this were to ionize in the reaction medium, then recombine, the observed product would be formed.

$$
\begin{array}{c}
CH_2-CH=C=CH_2 \\
| \\
Cl
\end{array}
\xrightarrow{-Cl^-}
\left[
\begin{array}{c}
\overset{+}{CH_2}-CH=C=CH_2 \\
\updownarrow \\
CH_2=CH-\overset{+}{C}=CH_2
\end{array}
\right]
\longrightarrow
$$

$$
\begin{array}{c}
CH_2=CH-C=CH_2 \\
| \\
Cl
\end{array}
$$

4-13.

$$
\left[CH_3C=CHCH_2 \mid CH_2C=CHCH \mid CH-C=CHCH \mid CH-C=CH-CH \right]_2
$$
$$
\hspace{1.2cm} | \hspace{1.7cm} | \hspace{2.0cm} | \hspace{2.2cm} |
$$
$$
\hspace{1.2cm} CH_3 \hspace{1.2cm} CH_3 \hspace{1.5cm} CH_3 \hspace{1.6cm} CH_3
$$

lycopene

carotene

(Note that at the central dashed line in the carotene structure, the two halves of the molecule are pieced together with isoprene units joined tail-to-tail; all other dashed lines show units joined head-to-tail).

4-14. The Cope rearrangement requires two double bonds separated by three sigma bonds (a total of six carbon atoms). To illustrate these, we have numbered the pertinent carbons in each structure 1-6 (these are *not* the numbers that would be used in naming the compounds).

Eq 4-29

Eq 4-30

CHAPTER FIVE

AROMATIC HYDROCARBONS

5-1. *a)*

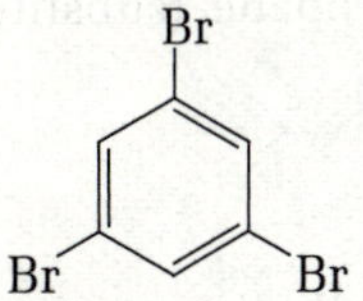

Br

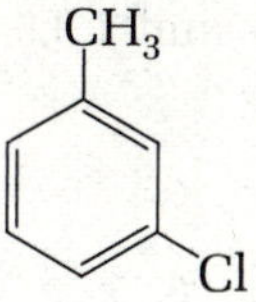

Br   Br

b) CH$_3$

Cl

c) CH$_2$CH$_3$

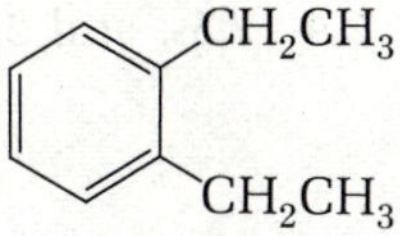

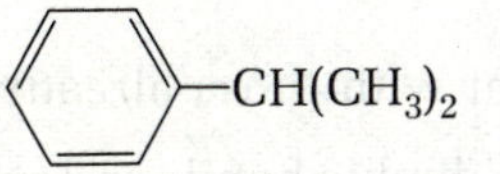

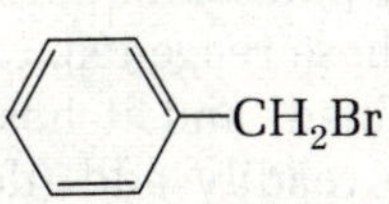

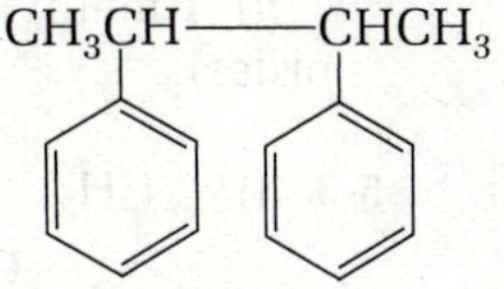

CH$_2$CH$_3$

d) —CH(CH$_3$)$_2$

e) —CH$_2$Br

f) CH$_3$CH————CHCH$_3$

g) Br————CH=CH$_2$

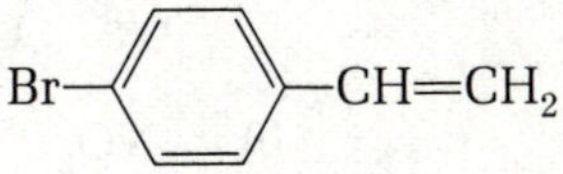

h)

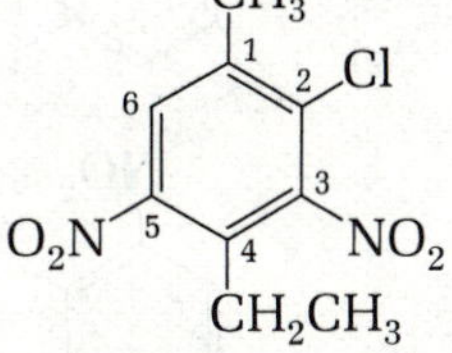

CH$_3$

1   2 Cl

6

3

O$_2$N 5   4 NO$_2$

CH$_2$CH$_3$

i)

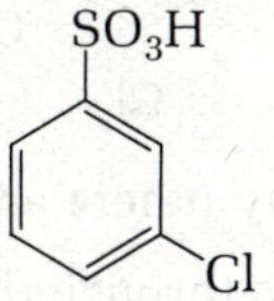

SO$_3$H

Cl

j) Br————CO$_2$H

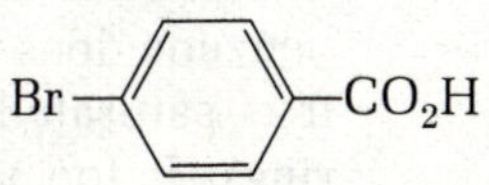

5-2. *a)* n-propylbenzene (or 1-phenylpropane)

b) o-chlorotoluene

c) m-bromochlorobenzene (alphabetic order)

d) 3,5-dibromostyrene

e) naphthalene

f) hexamethylbenzene (No numbers are necessary, since all possible positions on the benzene ring are substituted.)

g) 2,5-dichlorotoluene; number as is shown in the following structure.

$$CH_3$$

Cl₂... CH₃ ... 1,2 ... 6 ... 3 ... 5 Cl ... 4

h) 1-methyl-1-phenylcyclopropane (substituents in alphabetic order)

5-3. *a*)

CH₃
CH₃
CH₃

1,2,3-trimethylbenzene
(also the 1,2,4- and 1,3,5-isomers)

b)

NO₂
Cl
Cl

2,3-dichloronitrobenzene
(also the 2,4-; 2,5-; and 2,6-isomers)

NO₂
Cl
Cl

3,4-dichloronitrobenzene
(also the 3,5-isomer)

Any others are redundant with those already given.

5-4. *a* through *d* all contain double bonds and would easily decolorize bromine or be oxidized by potassium permanganate (whereas benzene does not react with these reagents). Compound *e*, though it is saturated, is highly strained, and it has two cyclopropane rings. It too would therefore readily add (decolorize) bromine; see equation 2-27.

5-5. The energy released on hydrogenating a carbon-carbon double bond is 26.5–30 kcal/mole (sec 5.2). With four double bonds, one would calculate that 106–120 kcal/mole should be liberated when cyclooctatetraene is hydrogenated. The observed value (110 kcal/mole) falls within this range and suggests that cyclooctatetraene has no appreciable resonance energy. Its tub-like shape (sec 5.7) prevents appreciable overlap of the *p* orbitals around the ring.

5-6. This method was used many years ago to distinguish between and determine the structures of *o*-, *m*- and *p*-isomers.

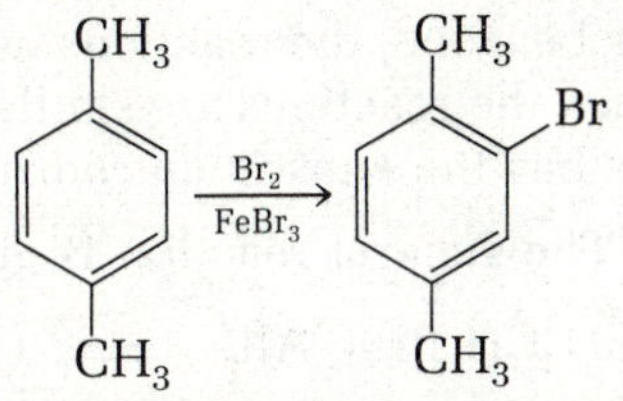

Only one isomer is possible, if the methyls are *para*.

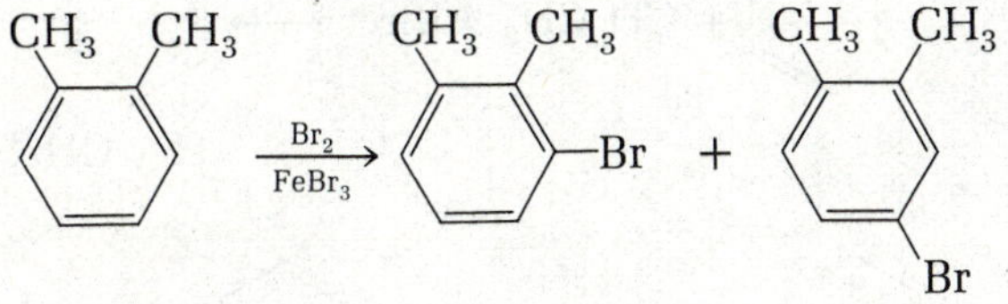

Two isomers are possible if the methyl groups are *ortho*.

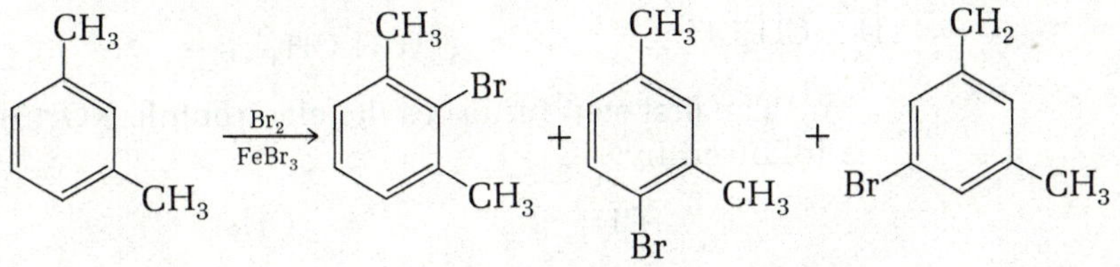

Three isomers are possible if the methyl groups are *meta*.

5-7. In each case, six carbons are required for the benzene ring; the remainder can be used for alkyl substituents.

a)

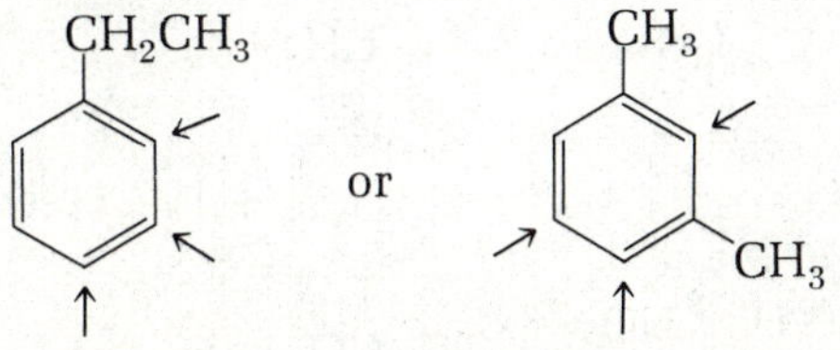

or

Both compounds would give three monobromo derivatives, as shown by the arrows.

b)

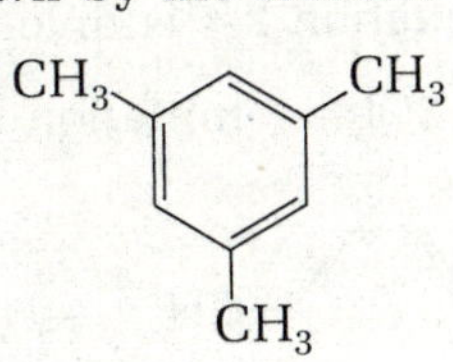

c)

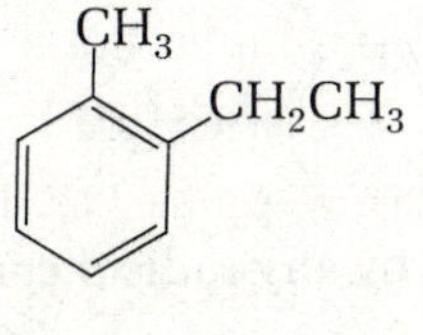

5-8. For benzene, the reactions are summarized in equation 5-8; for toluene, the reactions are similar, and the product which predominates has the substituent *para* to the methyl group.

5-9. *a*) The general reaction is given in equation 5-8.

$$CH_3CH_2Br + AlBr_3 \rightleftharpoons CH_3CH_2^+AlBr_4^- \qquad (eq\ 5\text{-}14)$$

$$+ CH_3CH_2^+AlBr_4^- \longrightarrow \qquad\qquad AlBr^- \qquad (eq\ 5\text{-}9)$$

$$\longrightarrow \qquad + HBr + AlBr_3 \qquad (eq\ 5\text{-}10)$$

b) The first step furnishes the electrophile NO_2^+ (eq 5-12). This is followed by:

$$+ {}^+NO_2 \longrightarrow \qquad\qquad (eq\ 5\text{-}9)$$

$$\longrightarrow \qquad + H^+ \qquad (eq\ 5\text{-}10)$$

c) The overall equation is 5-18. A free radical mechanism analogous to equation 2-2 to equation 2-4 is involved:

$$Cl_2 \xrightarrow[\text{light}]{uv} 2Cl\cdot \qquad \text{initiation step}$$

$$-CH_3 + Cl\cdot \longrightarrow \qquad -CH_2\cdot + HCl$$

$$-CH_2\cdot + Cl_2 \longrightarrow \qquad -CH_2Cl + Cl\cdot$$

free radical chain

The chain may be terminated by any radical combination reaction.

5-10. See section 5.4b for a discussion of the orienting influence of substituents.

a) 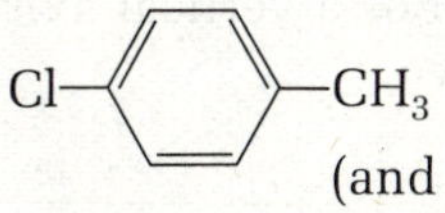 Cl—⟨benzene ring⟩—CH$_3$

(and *ortho*)

b) 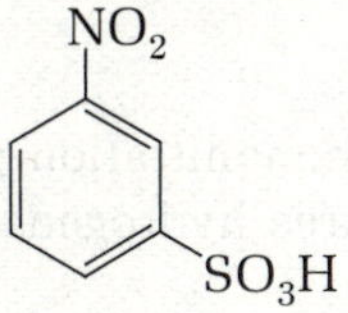 NO$_2$ ⟨benzene ring⟩ SO$_3$H

c) 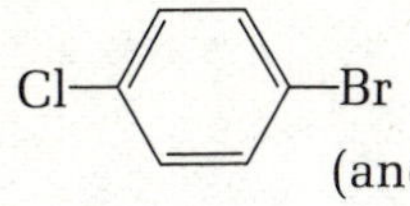 Cl—⟨benzene ring⟩—Br

(and *ortho*)

d) Same as *b.*

e) same as *c.*

f) 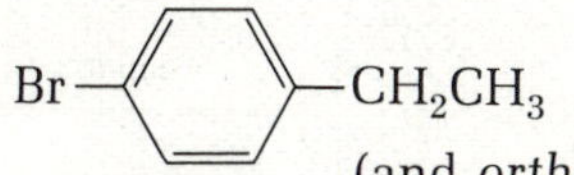 Br—⟨benzene ring⟩—CH$_2$CH$_3$

(and *ortho*)

g) 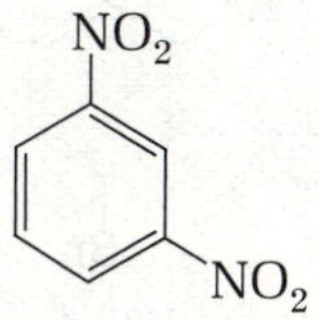 NO$_2$ ⟨benzene ring⟩ NO$_2$

h) 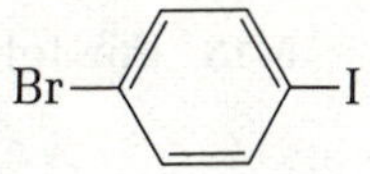 Br—⟨benzene ring⟩—I

5-11. Compare the two possible intermediates:

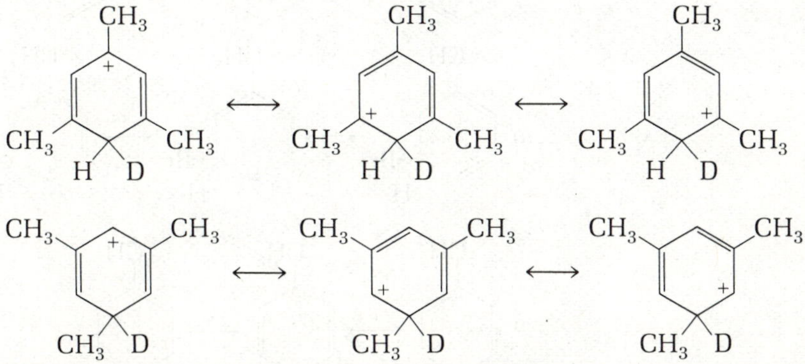

In the first benzenonium ion, the positive charge is stabilized, in all three resonance contributors, by an adjacent methyl group. Since tertiary carbonium ions are more stable than secondary carbonium ions, this is the preferred intermediate.

5-12. This problem is similar to 5-6, but involves starting with tri-, rather than disubstituted benzenes.

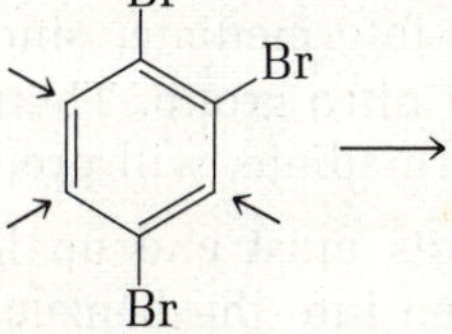

Three mononitration products are possible, as shown by the arrows.

mp 44°

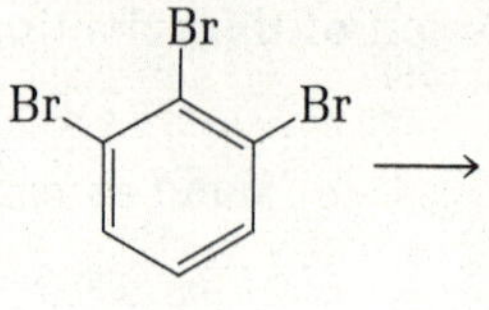

mp 87°

Two mononitration products are possible ⟶ because of a vertical axis of symmetry, as drawn.

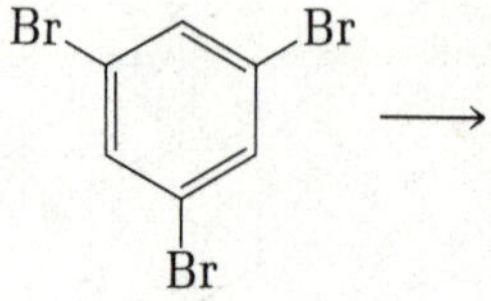

mp 119°

Only one mononitration product is possible, ⟶ since all three hydrogens are equivalent.

5-13. For toluene:

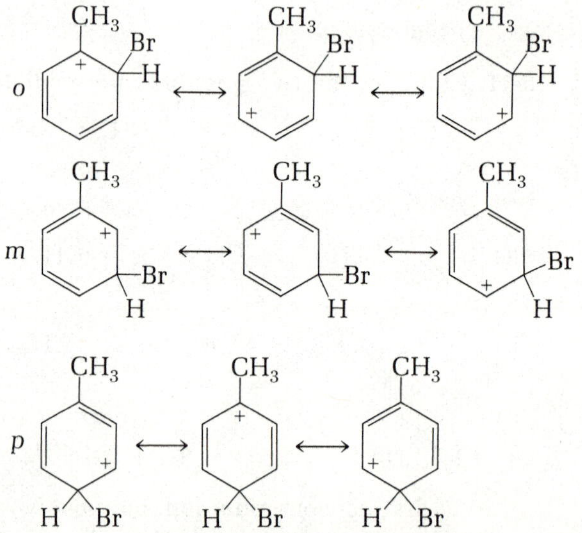

Only the o- and p-intermediates can have the positive charge adjacent to and stabilized by the methyl group; products derived from these intermediates therefore predominate.

For nitrobenzene: Draw the same structures as for toluene, but replace the CH_3 by NO_2. Since the NO_2 group is electron-withdrawing it destabilizes carbonium ions. The ion which will be least destabilized is the m-intermediate, since the positive charge is never adjacent to the nitro group. Therefore the *meta* product, derived from this intermediate, will predominate.

5-14. *a*) Since the two substituents must end up in a *meta* relationship, the first one introduced into the benzene ring must be *meta*-directing. Therefore nitrate first, then brominate.

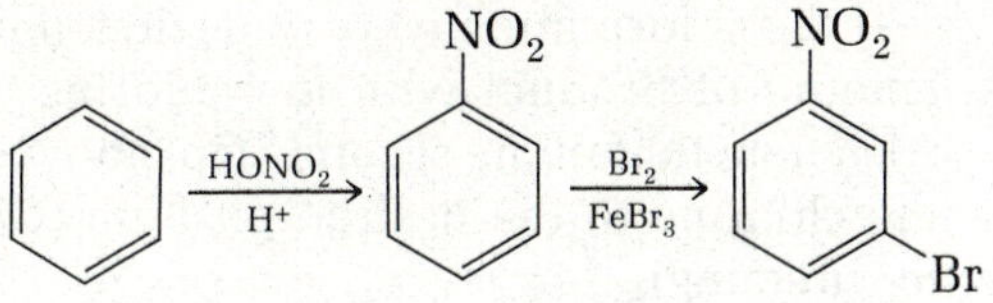

b)

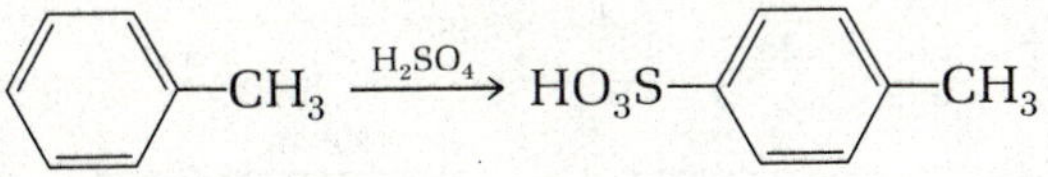

Alkyl groups are *o,p*-directing.

c)

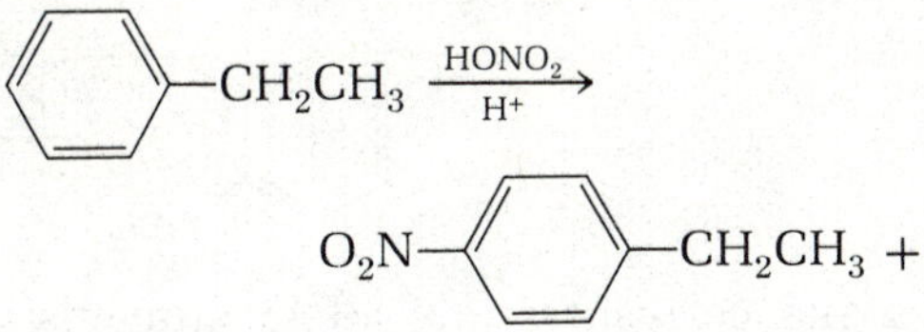

O_2N- $-CH_2CH_3$ + some *ortho* isomer

d) 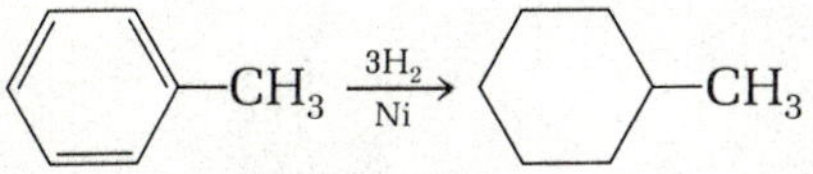 (eq 5-17)

e)

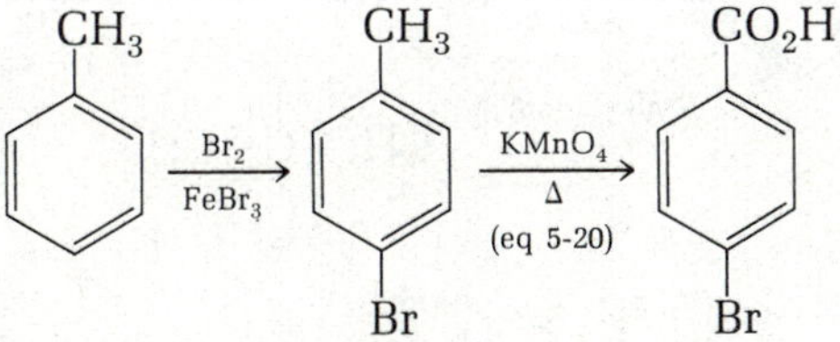

f) equations 5-25 and 5-26.

g) Compare with part *a*. The bromination must be performed first.

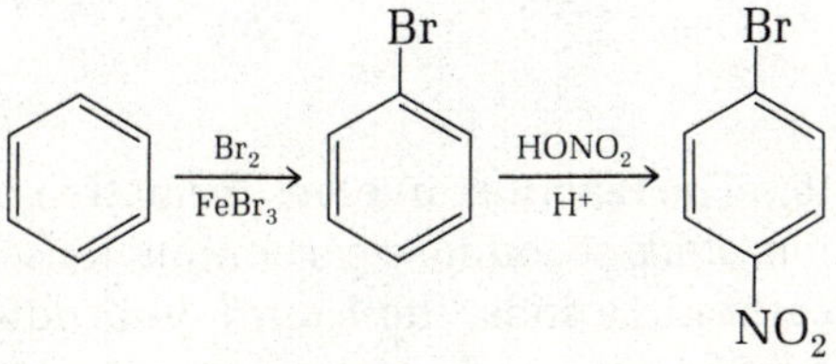

h)

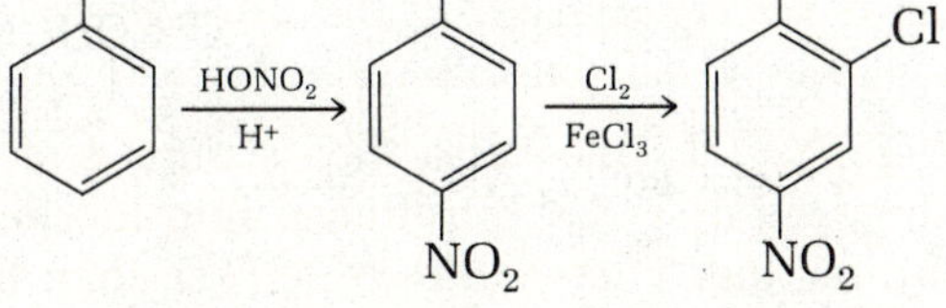

If the chlorination were done first, one would get a considerable amount of product with the chlorine *para* to the methyl group. Also note that in the second step, above, both substituents "direct" the chlorine to the desired position (CH_3 is *o,p*-directing, NO_2 is *m*-directing).

i)

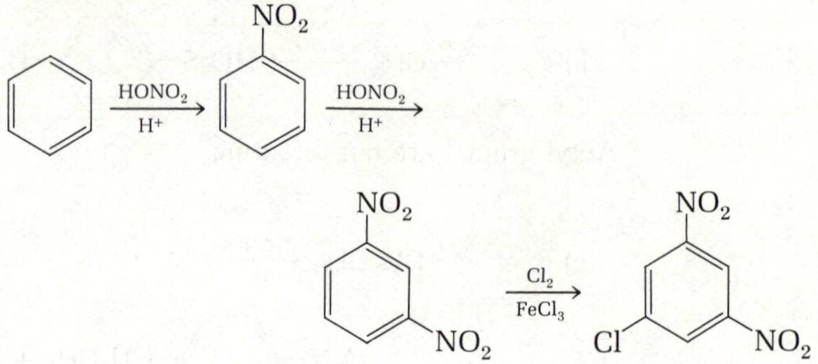

This sequence insures that all substituents will be *meta* to one another.

j)

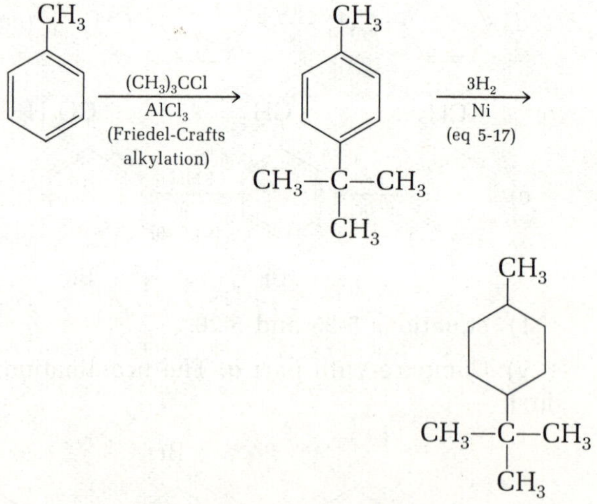

5-15. The reaction involves formation of an intermediate carbonium ion; since secondary carbonium ions are more stable than primary carbonium ions, the isopropyl product is formed.

$$CH_3CH{=}CH_2 \xrightarrow[AlCl_3]{HCl} CH_3\overset{+}{C}HCH_3\ AlCl_4^-$$

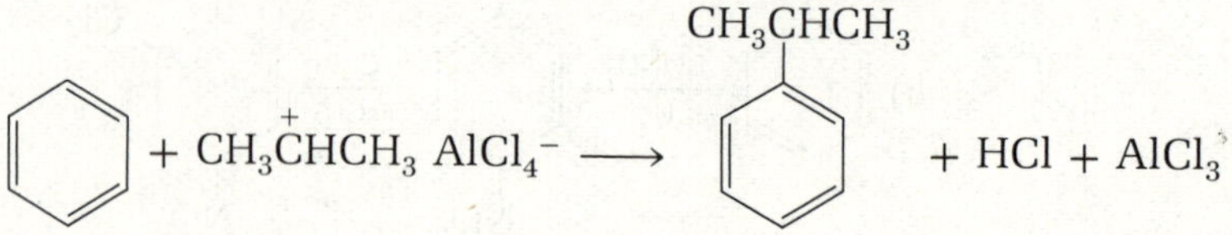

(See also Markownikoff's rule, sec 3.4c, especially eq 3-18).

5-16. The nitro group has two main contributing structures:

$$-\overset{+}{N}\overset{\ddot{\ddot{O}}}{\underset{\ddot{\ddot{O}}:}{\diagdown}} \quad \longleftrightarrow \quad -\overset{+}{N}\overset{\ddot{O}:^-}{\underset{\diagdown O:}{}}$$

Since they are identical and contribute equally, there is only one type of N—O bond, intermediate between double and single in length.

5-17. *a*) Since the carboxyl group is $-CO_2H$, the acid has the formula $C_6H_5CO_2H \ (\equiv C_7H_6O_2)$. This must be benzoic acid. Therefore the original hydrocarbon C_9H_{12} which was oxidized to this acid (eq 5-20 to 5-22) can have only one alkyl substituent. Possibilities are

$-CH_2CH_2CH_3$ and $-CH(CH_3)_2$

b) The original hydrocarbon must have two alkyl substituents *ortho* to one another; the only possible structure is

$\xrightarrow{\text{oxidize}}$ (cf. eq 5-22)

o-ethyltoluene   o-phthalic acid

5-18. The mechanism is analogous to that given for polyethylene; review section 3.4d, especially equations 3-20 to 3-22.

$$R\cdot \ + \ CH_2{=}CH \ \longrightarrow \ R{-}CH_2{-}\overset{\cdot}{C}H \qquad \text{initiation}$$

from a
radical source

$$R{-}CH_2{-}\overset{\cdot}{C}H \ + \ CH_2{=}CH \ \longrightarrow \ RCH_2\,CH{-}CH_2\overset{\cdot}{C}H \qquad \text{propagation}$$

Termination may be by radical combination, or by loss of an adjacent H atom to leave one double bond at the end of the chain.

Note that the radicals add to styrene in such a way that the new radical is always adjacent to the aromatic ring, where it can be

stabilized by resonance. That is,

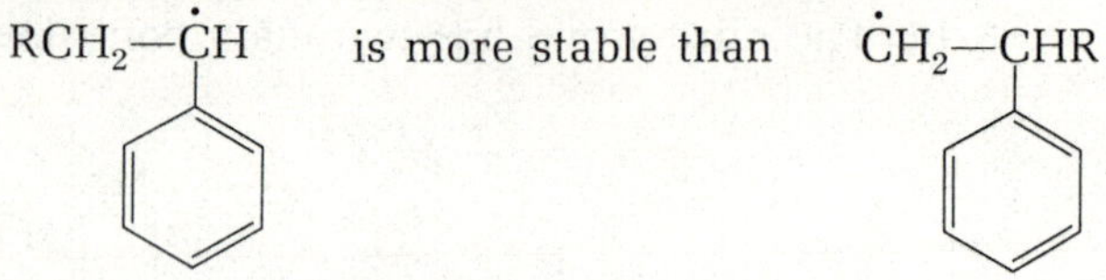

RCH$_2$—ĊH   is more stable than   ĊH$_2$—CHR

because of the delocalization as shown:

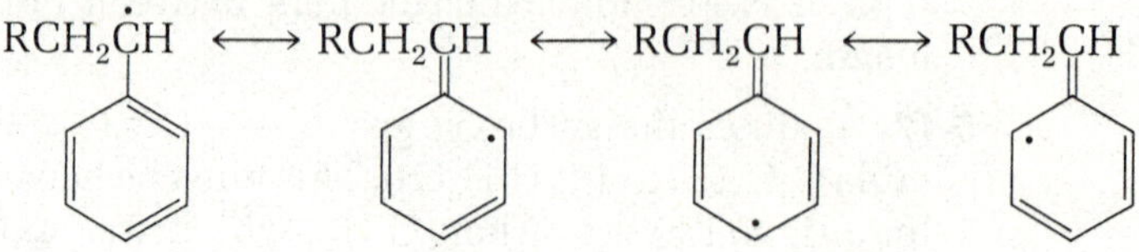

RCH$_2$ĊH ⟷ RCH$_2$CH ⟷ RCH$_2$CH ⟷ RCH$_2$CH

5-19. 1-bromonaphthalene:

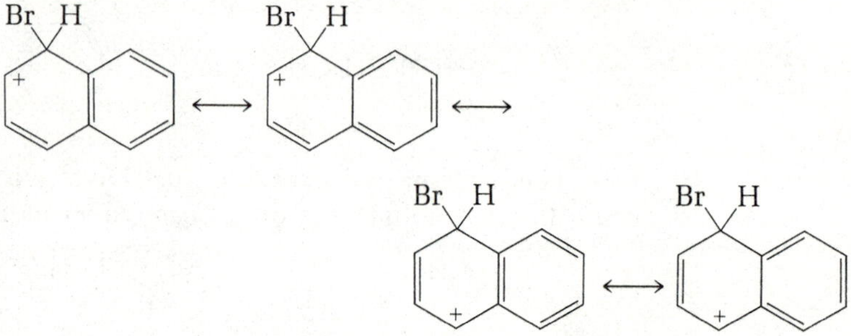

2-bromonaphthalene:

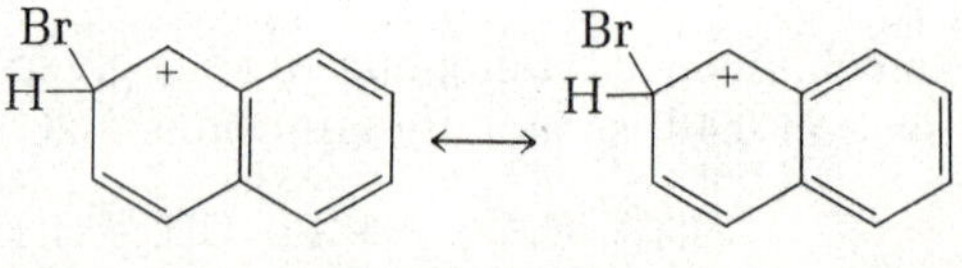

Clearly substitution in the 1 position is preferred, since the charge in the intermediate can be more extensively delocalized. Other resonance contributors can be written which destroy the aromaticity of the second ring; even if these are included, 1-substitution will be preferred.

5-20.

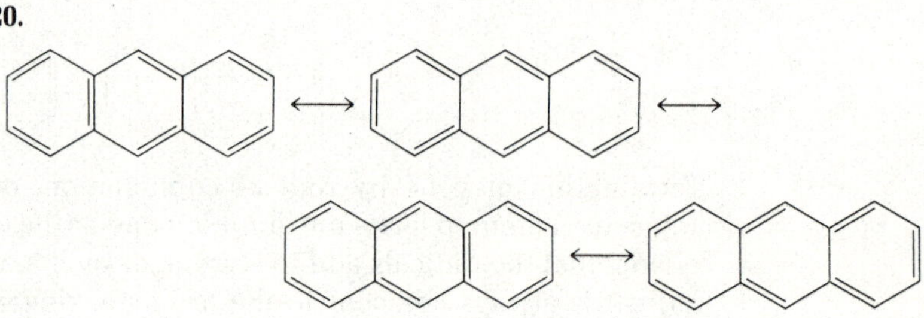

In these structures, either two rings are aromatic (three double bonds) and one is not (only two double bonds), or one ring is aromatic and two are not. If anthracene is reduced at the 9,10-positions, two fully aromatic rings of the benzene type remain.

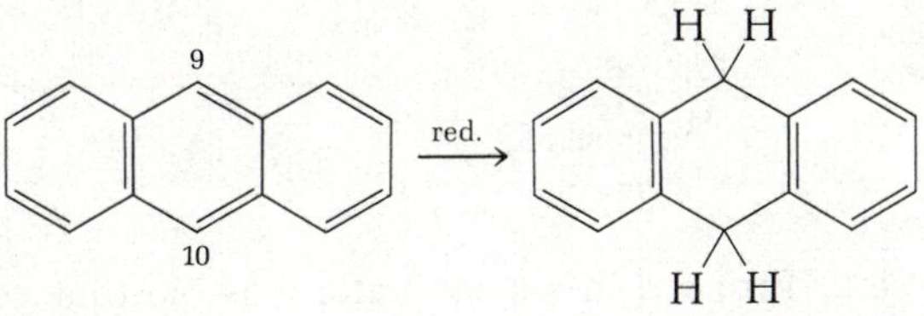

9,10-dihydroanthracene

Very little resonance energy is lost in this process, but further reduction would be nearly as difficult as for benzene itself.

Other dihydroanthracenes are possible, as for example 1,4-dihydroanthracene, but this would lead to a compound with an isolated double bond which would easily be further reduced.

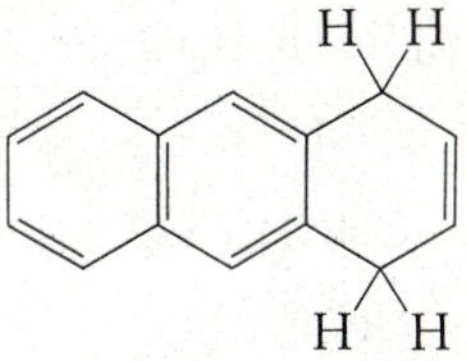

CHAPTER SIX

ALCOHOLS AND PHENOLS

6-1. a) $CH_2CCH_2CH_3$ with CH_3 above and OH, CH_3 below

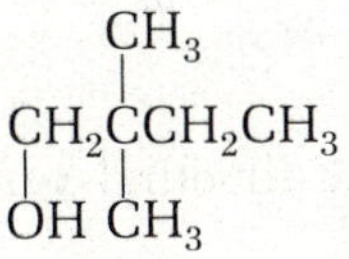

b) OH, Br on ring

c) $CH_3CH\ CHCH_2CH_3$ with $OH\ OH$ below

d) ring—CH_2CHCH_2—ring with OH below

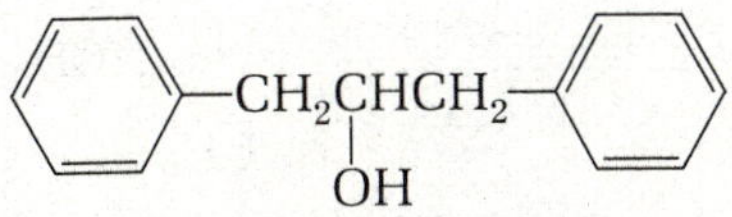

e) $CH_3CH_2CH_2OSO_3H$

f) $\overset{1}{C}H_3\overset{2}{C}H\overset{3}{C}H=\overset{4}{C}H\overset{5}{C}H_3$ with OH below

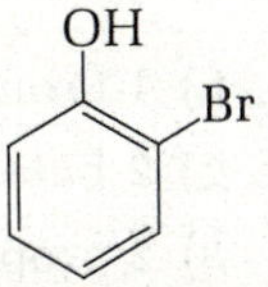

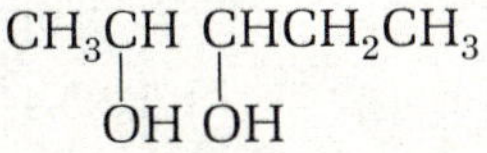

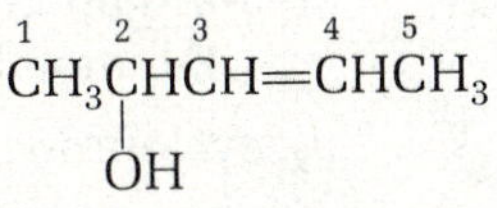

g) 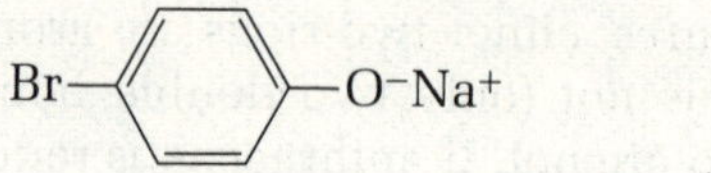 Br—⟨benzene ring⟩—O⁻Na⁺

i) 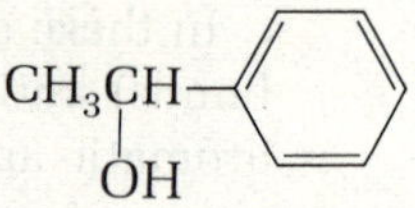 CH_3CH—⟨benzene ring⟩
$\quad\quad\quad |$
$\quad\quad\quad OH$

h) HO   H
   ⟨cyclopentane ring with positions 1,2,3,4,5⟩
   $\quad\quad\quad$—CH_3
   H

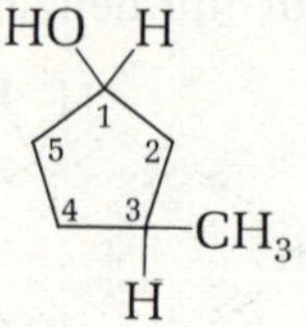

j) 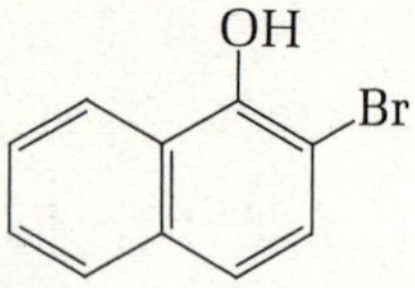 ⟨naphthalene ring⟩
   OH
   Br

6-2. Proceed in a systematic way, so that you don't miss any of the possibilities.

$CH_2CH_2CH_2CH_2CH_3$ 1-pentanol (1°)
$|$
OH

$CH_3CHCH_2CH_2CH_3$ 2-pentanol (2°)
$\quad\,|$
$\quad OH$

$CH_3CH_2CHCH_2CH_3$ 3-pentanol (2°)
$\quad\quad\,|$
$\quad\quad OH$

CH_2—$CHCH_2CH_3$ 2-methyl-1-butanol (1°)
$|\quad\quad|$
$OH\quad CH_3$

$\quad\quad OH$
$\quad\quad |$
$CH_3CCH_2CH_3$ 2-methyl-2-butanol (3°)
$\quad\quad |$
$\quad\quad CH_3$

CH_3CH——$CHCH_3$ 3-methyl-2-butanol (2°)
$\quad\,|\quad\quad\,|$
$\quad CH_3\quad OH$

$CH_3CHCH_2CH_2OH$ 3-methyl-1-butanol (1°)
$\quad\,|$
$\quad CH_3$

$\quad\quad\,CH_3$
$\quad\quad\,|$
CH_2—C—CH_3 2,2-dimethyl-1-propanol (1°)
$|\quad\quad|$
$OH\quad CH_3$

6-3. a) 2-butanol

b) 2,4-dichlorophenol

c) 3-bromo-2-methyl-2-butanol

d) cyclopropanol

e) m-bromophenol

f) 1-bromo-2-naphthol

g) 2-buten-1-ol

h) 2-propanethiol (or iso-propyl mercaptan)

i) 1,2,3,4-butanetetraol

j) cis-3-methyl-cyclobutanol

6-4. *a*) This name is frequently used, yet is incorrect since it mixes two nomenclature systems. The correct names should be either isopropyl alcohol (the common name) or 2-propanol (the IUPAC name).

b) The hydroxyl should get the lower number; 3,3-dimethyl-2-butanol.

c) 2-methyl-1-butanol; the longest chain was not selected.

d) 2-propen-1-ol (or allyl alcohol); the hydroxyl group should get the lower number.

e) 3-chlorocyclohexanol; number the ring from the hydroxyl-bearing carbon, in a direction that gives substituents the lowest possible numbers.

f) phenol; the term alcohol is not used when the hydroxyl group is on an aromatic ring.

g) 2-bromo-*p*-cresol or 2-bromo-4-methylphenol; give substituents the lowest possible numbers.

h) 1,2-propanediol; the chain was numbered in the wrong direction.

6-5. *a*) ethyl chloride < 1-hexanol < ethanol. Both alcohols can hydrogen-bond to water and will be more soluble than the alkyl chloride. The lower molecular weight alcohol will be more soluble (it has a shorter hydrophobic carbon chain).

b) The order given is correct; the more hydroxyl groups, the more possibilities for hydrogen bonding and the greater the water solubility.

c) benzene < phenol < hydroquinone < sodium phenoxide. The hydrocarbon is nonpolar; hydroquinone has two hydroxyl groups and is expected to be more soluble than phenol, which has only one; the phenoxide is an ionic compound and would be expected to be most soluble of all.

6-6. Boiling Point: dimethyl ether < methanol < water
Molecular Weight: H_2O < CH_3OH < CH_3OCH_3
In the absence of other factors, boiling point and molecular weight usually parallel one another; the greater the molecular weight the greater the number of atoms and electrons per molecule, the greater the intermolecular attractive forces and the greater the energy needed to separate molecules one from another—hence, the greater the boiling point.

The other factor, which reverses the order in the example being discussed, is hydrogen bonding. This is greatest for water, less significant for methanol, and unimportant with dimethyl ether. Review section 6.3.

6-7. One might expect CH_3SH to have a higher boiling point than CH_3OH if molecular weight were the only factor (48 vs. 32). But hydrogen bonding is also important, and is expected to be more important for CH_3OH than for CH_3SH, since oxygen is considerably more electronegative than sulfur (sec 1.2). Hydrogen bonding is the dominant factor. The actual boiling points are: CH_3OH, 65°; CH_3SH, 7°.

6-8. The term amphoteric means that a substance can act either as an acid or a base. Perhaps the simplest equation which illustrates alcohols acting in both capacities is their autoprotolysis (eq 6-7).

$$R-\overset{..}{\underset{..}{O}}-H + R-\overset{..}{\underset{..}{O}}-H \rightleftharpoons \left[R-\overset{H}{\underset{..}{\overset{..}{O}}}-H\right]^+ + R-\overset{..}{\underset{..}{O}}:^-$$

acts as a base; accepts a proton.   acts as an acid; donates a proton

Review sec 6.4.

6.9. a) cyclohexanol $<$ p-cresol $<$ phenol $<$ p-chlorophenol. Phenols are much more acidic than alcohols for reasons discussed in section 6-4. Therefore cyclohexanol would be the least acidic of the four. Amongst the three phenols, acidity is enhanced by electron-withdrawing substituents; since Cl is electron-withdrawing and CH_3 is electron-donating, relative to hydrogen, the expected order would be as shown.

b)

The negative charge is spread over the three oxygens (as well as other positions). Since this is not possible for phenoxide ion itself, one expects p-nitrophenol to be a stronger acid than phenol. In fact, the K_a's are 1.1×10^{-10} for phenol and 6.9×10^{-8} for p-nitrophenol (it is the stronger acid by a factor of more than 600).

6-10. a) $CH_2{=}\underset{CH_3}{\overset{|}{C}}{-}CH_3 + H_2O \xrightarrow{H^+} CH_3\underset{CH_3}{\overset{OH}{\overset{|}{\underset{|}{C}}}}CH_3$ (eq 6-11, 6-12)

b) —CH=CH$_2$ + H$_2$O $\xrightarrow{\text{H}^+}$ —CHCH$_3$
                                                                            |
                                                                           OH

(eq 6-11, 6-12)

Both *a* and *b* involve Markownikoff addition to the alkene.

c) —CH=CH$_2$ + (BH$_3$)$_2$ $\longrightarrow$ $\left(\text{} -CH_2-CH_2 \right)_3$B

—CH$_2$CH$_2$OH $\xleftarrow{\text{OH}^-}$ $\left(\text{} -CH_2CH_2O \right)_3$B $\xleftarrow{\text{H}_2\text{O}_2}$

This is an example of hydroboration (sec 6.5b).

d) CH$_3$CH$_2$CH$_2$CH$_2$Br $\xrightarrow[\text{OH}^-]{\text{aqueous}}$

CH$_3$CH$_2$CH$_2$CH$_2$OH   (sec 6.5c, eq 6-16).

e) CH$_3$CH=CH$_2$ $\xrightarrow[\text{heat}]{\text{Cl}_2,}$ CH$_2$CH=CH$_2$ $\xrightarrow[\text{OH}^-]{\text{aqueous}}$ CH$_2$CH=CH$_2$
                                                | |
  eq 3-27a   Cl   eq 6-16   OH

f) $\xrightarrow[\substack{\text{and alkali}\\ \text{fusion}\\ \text{(sec 6.6a)}}]{\text{sulfonation}}$ $\xrightarrow[\text{Ni}]{\text{H}_2}$
  eq 6-39

g) $\xrightarrow[\text{heat}]{\text{H}_2\text{SO}_4}$ $\xrightarrow[\text{heat}]{\text{H}_2\text{SO}_4}$ $\xrightarrow{\text{NaOH}}$

$\xrightarrow[300°]{\text{heat}}$

This is an extension of the alkali fusion of sulfonates, section 6.6a. Since the SO$_3$H group is *meta*-directing, the second sulfonation product is *meta*-benzenedisulfonic acid.

h)

$$\underset{\substack{\text{nitrate,}\\\text{then sulfonate}\\\text{(or vice versa)}}}{\longrightarrow} \quad \text{(SO}_3\text{H, NO}_2 \text{ benzene)} \quad \underset{\substack{\text{NaOH}\\\text{fusion}\\\text{(sec 6.6a)}}}{\longrightarrow} \quad \text{(OH, NO}_2 \text{ benzene)}$$

Since both the NO_2 and SO_3H groups are *meta*-directing, the sequence selected to obtain *m*-nitrobenzenesulfonic acid from benzene is not too important.

i) $CH_3CH{=}CHCH_3 \xrightarrow{\text{HOCl}} CH_3\underset{\overset{|}{OH}}{CH}{-}\underset{\overset{|}{Cl}}{CH}CH_3 \xrightarrow[\substack{Na_2CO_3 \text{ in}\\ H_2O}]{OH^-}$

$$CH_3\underset{\overset{|}{OH}}{CH}{-}\underset{\overset{|}{OH}}{CH}CH_3$$

This is an extension of the method used for ethylene glycol, equation 6-40.

j) $\text{(benzene)}{-}CH_3 \xrightarrow[\substack{\text{ultraviolet}\\\text{light}\\(\text{eq 5-18})}]{Cl_2} \text{(benzene)}{-}CH_2Cl \xrightarrow[\substack{\text{aqueous}\\\text{base}\\(\text{eq 6-16})}]{\text{dil.}}$

$$\text{(benzene)}{-}CH_2OH$$

6-11. a) $(CH_3)_3COH + HCl \longrightarrow (CH_3)_3CCl + H_2O$ (eq 6-27)

b) $2\ CH_3CH_2CH_2CH_2CH_2OH + 2\ Na \longrightarrow$
$$2\ CH_3CH_2CH_2CH_2CH_2O^-Na^+ + H_2$$
sodium 1-pentoxide   (eq 6-4)

c) $\text{(benzene)}{-}CH_2Cl + Na^+OH^- \xrightarrow{H_2O}$

$$\text{(benzene)}{-}CH_2OH + Na^+Cl^- \quad (\text{eq 6-16})$$

d) $CH_3{-}\text{(benzene)}{-}OH + 3H_2 \xrightarrow[\text{heat}]{Ni} \text{(cyclohexane with }CH_3, H, H, OH\text{)}$

mixture of *cis*
and *trans* isomers   (eq 6-39)

e) $CH_2{=}CHCH_2OH + HOCl \longrightarrow$

$$\underset{\overset{|}{Cl}}{CH_2}{-}\underset{\overset{|}{OH}}{CH}{-}CH_2OH \quad (\text{eq 6-42})$$

f) 3 $+ PBr_3 \longrightarrow 3$ $+ H_3PO_3$ (eq 6-31)

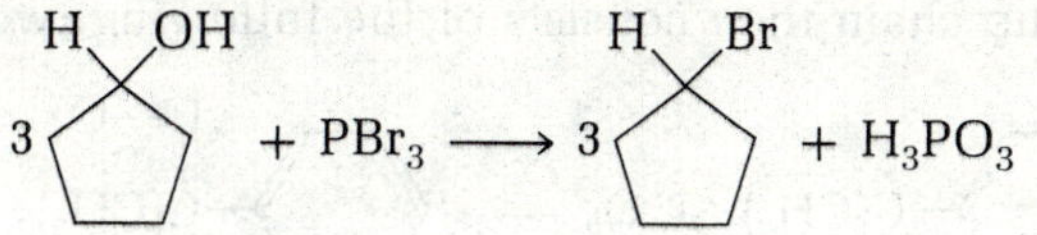

g) CH_3CH- $+ SOCl_2 \longrightarrow$

CH_3CH- $+ SO_2 + HCl$ (eq 6-30)

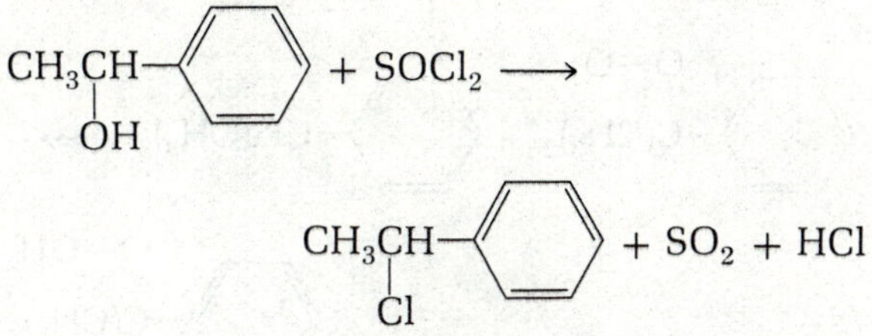

h) $CH_3CH_2CH_2CH_2OH + HOSO_3H \xrightarrow{\text{cold}}$

$CH_3CH_2CH_2CH_2OSO_3H + H_2O$ (eq 6-33)

i) $+ Na^+OH^- \xrightarrow{H_2O}$ no reaction   (sec 6.4)

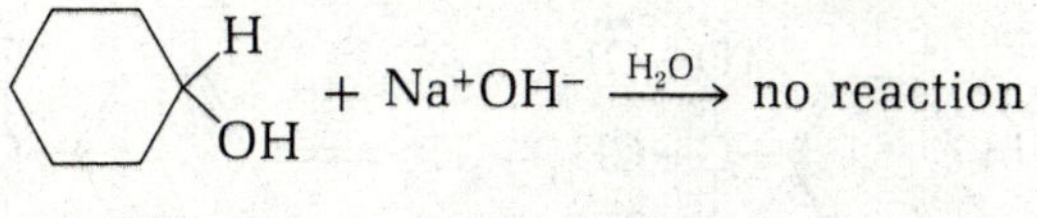

j) $+ Na^+OH^- \longrightarrow$

$+ H_2O$ (eq 6-9)

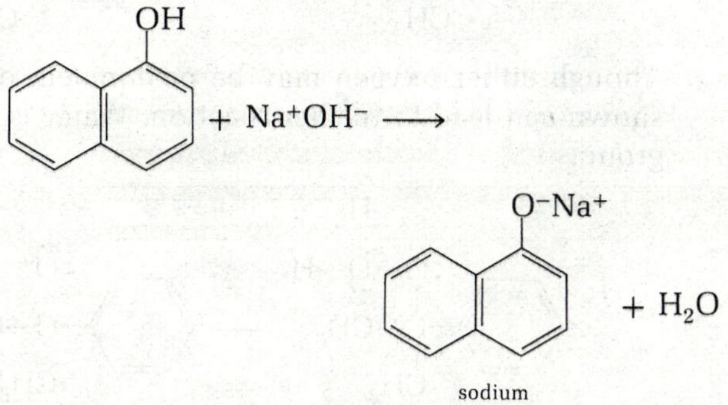

sodium
1-naphthoxide

6-12. *a)* This is a free radical chain reaction, initiated by some radical R.

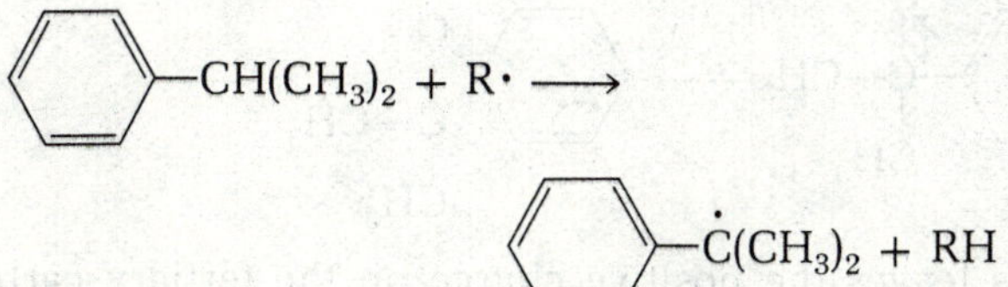

$-CH(CH_3)_2 + R\cdot \longrightarrow$

$-\dot{C}(CH_3)_2 + RH$ (initiation step)

The hydrogen adjacent to the aromatic ring is abstracted preferentially, since the resulting radical is resonance-stabilized:

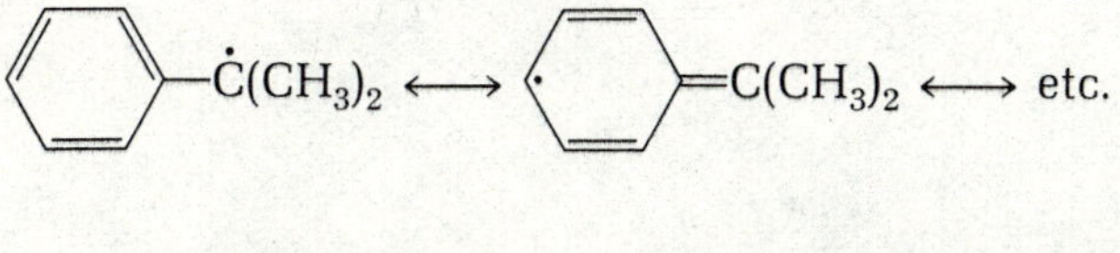

$-\dot{C}(CH_3)_2 \longleftrightarrow \cdot$ $=C(CH_3)_2 \longleftrightarrow$ etc.

The chain then consists of the following two steps:

$$\text{Ph}-\overset{\cdot}{\text{C}}(\text{CH}_3)_2 + \text{O}_2 \longrightarrow \text{Ph}-\underset{\text{C}(\text{CH}_3)_2}{\overset{\text{O}-\text{O}\cdot}{|}}$$

$$\text{Ph}-\underset{\text{C}(\text{CH}_3)_2}{\overset{\text{O}-\text{O}\cdot}{|}} + \text{Ph}-\text{CH}(\text{CH}_3)_2 \longrightarrow$$

$$\text{Ph}-\underset{\text{C}(\text{CH}_3)_2}{\overset{\text{O}-\text{OH}}{|}} + \text{Ph}-\overset{\cdot}{\text{C}}(\text{CH}_3)_2$$

Any radical combination will terminate the chain.

b)
$$\text{Ph}-\underset{\overset{|}{\text{CH}_3}}{\overset{:\ddot{\text{O}}-\ddot{\text{O}}\text{H}}{\underset{}{\text{C}}-\text{CH}_3}} + \text{H}^+ \rightleftharpoons \text{Ph}-\underset{\overset{|}{\text{CH}_3}}{\overset{:\ddot{\text{O}}-\overset{\text{H}}{\underset{}{\overset{|}{\text{O}}}}{}^{+}\text{H}}{\underset{}{\text{C}}-\text{CH}_3}}$$

Though either oxygen may be protonated, only the protonation shown can lead to further reaction. Water is now a good leaving group:

$$\text{Ph}-\underset{\overset{|}{\text{CH}_3}}{\overset{:\ddot{\text{O}}-\overset{\text{H}}{\underset{}{\overset{|}{\text{O}}}}{}^{\pm}\text{H}}{\underset{}{\text{C}}-\text{CH}_3}} \longrightarrow \text{Ph}-\underset{\overset{|}{\text{CH}_3}}{\overset{:\ddot{\text{O}}^{+}}{\underset{}{\text{C}}-\text{CH}_3}} + \text{H}_2\text{O}$$

The phenyl now "migrates" to the positive oxygen. The reaction can be viewed as an intramolecular electrophilic aromatic substitution:

$$\longrightarrow \longrightarrow \text{Ph}-\ddot{\text{O}}-\overset{+}{\underset{\overset{}{\text{CH}_3}}{\overset{\text{CH}_3}{\text{C}}}}$$

This leaves the positive charge on the tertiary carbon, where it can also be stabilized by delocalization to the oxygen:

$$\text{Ph}-\ddot{\text{O}}-\overset{+}{\underset{\overset{}{\text{CH}_3}}{\overset{\text{CH}_3}{\text{C}}}} \longleftrightarrow \text{Ph}-\overset{+}{\ddot{\text{O}}}=\underset{\overset{}{\text{CH}_3}}{\overset{\text{CH}_3}{\text{C}}}$$

Attack on the tertiary carbonium ion by the nucleophilic solvent (water) completes the process:

$$\text{C}_6\text{H}_5-\overset{+}{\text{O}}-\overset{\overset{\displaystyle CH_3}{|}}{\underset{\underset{\displaystyle CH_3}{|}}{C}}-CH_3 + H_2O \longrightarrow \text{C}_6\text{H}_5-\overset{..}{\underset{..}{O}}-\overset{\overset{\displaystyle CH_3}{|}}{\underset{\underset{\displaystyle \overset{+}{O}}{|}}{C}}-CH_3$$

$$\overset{\overset{\displaystyle CH_3}{|}}{\underset{\underset{\displaystyle \overset{+}{O}}{|}}{C}} \quad \overset{H\quad H}{}$$

$$\text{C}_6\text{H}_5-O-\overset{\overset{\displaystyle CH_3}{|}}{\underset{\underset{\displaystyle \overset{+}{O}}{|}}{C}}-CH_3 \rightleftharpoons \text{C}_6\text{H}_5-\overset{+}{O}-\overset{\overset{\displaystyle H\quad CH_3}{|\quad|}}{\underset{\underset{\displaystyle O-H}{}}{C}}-CH_3 \longrightarrow$$

$$\text{C}_6\text{H}_5-OH + (CH_3)_2C{=}O + H^+$$

6-13. In the laboratory all separations of this type are usually performed by dissolving the mixture in a low-boiling, inert organic solvent such as ether or methylene chloride. The solution is then extracted with an aqueous (neutral, acidic, or alkaline) solution which will extract one of the two components. The layers are then separated, and the organic layer is evaporated to recover the compound which was *not* extracted into the aqueous phase. The aqueous layer is then treated in some way to recover the extracted compound.

a) $\xrightarrow[\text{layer}]{\text{upper}}$ benzene (in ether) $\xrightarrow[\text{the ether}]{\text{evaporate}}$ benzene

benzene + phenol

$\left.\begin{array}{l}\text{aqueous} \\ \text{Na}^+\text{OH}^-\end{array}\right.$

(dissolved in an inert solvent, such as ether)

$\xrightarrow[\text{layer}]{\text{lower}}$ C$_6$H$_5$-O$^-$Na$^+$

in water

1) acidify with HCl
2) extract with ether

$\xrightarrow[\text{layer}]{\text{upper}}$ C$_6$H$_5$-OH

in ether (evaporate the ether to recover the phenol)

$\xrightarrow[\text{layer}]{\text{lower}}$ H$_2$O, Na$^+$Cl$^-$, excess HCl

(discard) this layer)

In the remaining answers, these procedures will be abbreviated.

b) The same procedure used in part *a* will work here, since phenol is extracted by base whereas cyclohexanol, a much weaker acid, is not.

c) An ether solution of the two alcohols can be extracted with water. 1-Propanol is soluble, whereas 1-heptanol, with a much longer carbon chain, is not.

6-14. *a*) Treat each with metallic sodium; bubbles of H_2 gas will be seen in the case of 3-pentanol, whereas the alkane is inert toward sodium (eq 6-4).

$$2-CH_3CH_2\underset{\underset{OH}{|}}{C}HCH_2CH_3 + 2\,Na \longrightarrow 2\,CH_3CH_2\underset{\underset{O^-Na^+}{|}}{C}HCH_2CH_3 + H_2$$

b) Treat each with a dilute solution of bromine in carbon tetrachloride; allyl alcohol, being unsaturated, will decolorize the bromine whereas isopropyl alcohol, being saturated, will not (eq 3-5).

$$CH_2{=}CHCH_2OH + Br_2 \longrightarrow \underset{\underset{Br}{|}}{C}H_2{-}\underset{\underset{Br}{|}}{C}H{-}CH_2OH$$

c) Add a little of each compound to 10% aqueous sodium hydroxide; the p-cresol will dissolve, whereas the benzyl alcohol is too weak an acid to react with aqueous base.

$$CH_3{-}\langle\!\bigcirc\!\rangle{-}OH + Na^+OH^- \longrightarrow CH_3{-}\langle\!\bigcirc\!\rangle{-}O^-Na^+ + H_2O$$

soluble in water

d) Smell the two compounds; thiols (mercaptans) have a strong, foul, unmistakable odor. This is a chemical test, since odor is the result of a chemical interaction (poorly understood) between a substance and the olfactory glands. One could also treat each compound with mercuric nitrate; the mercaptan would give an insoluble mercaptide (eq 6-45), whereas the alcohol would not react.

$$2\,CH_3CH_2CH_2CH_2SH + Hg^{+2}(NO_3)_2{}^{-1} \longrightarrow$$
$$(CH_3CH_2CH_2CH_2S)_2Hg + 2\,H^+NO_3^-$$

6-15. Work backward from the last data. The precursor of cyclohexene must be cyclohexyl bromide. This, in turn, must have come from cyclohexanol.

A   B   cyclohexene

The first reaction referred to is:

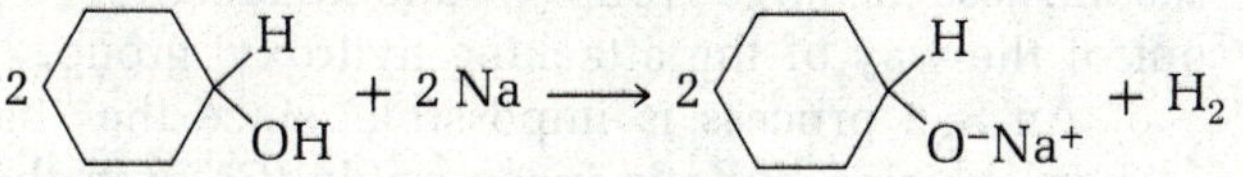

$$2 \overset{H}{\underset{OH}{\bigcirc}} + 2\,Na \longrightarrow 2 \overset{H}{\underset{O^-Na^+}{\bigcirc}} + H_2 \qquad (eq\ 6\text{-}4)$$

6-16. The ratio of C to H suggests immediately that the compound is aromatic. The solubility in dilute base suggests a phenol. Since there are seven carbons, the possibilities are:

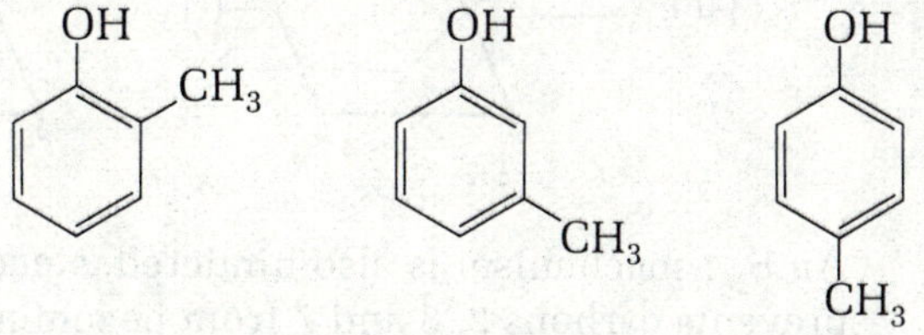

Reaction with bromine water to give a dibromo derivative (see eq 6-38) suggests that the compound is either *o*- or *p*-cresol (the *meta* isomer should give a tribromo derivative).

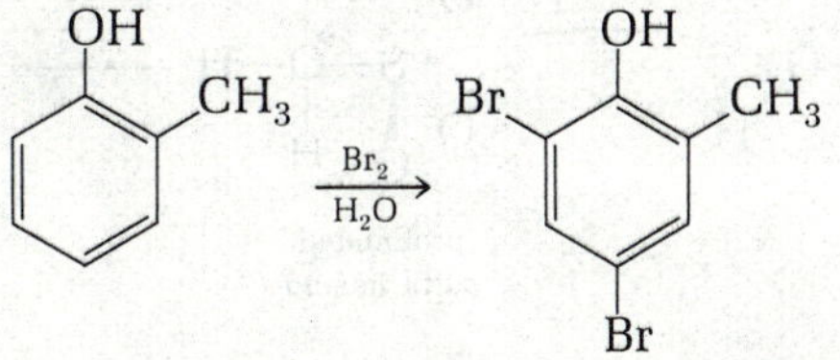

or

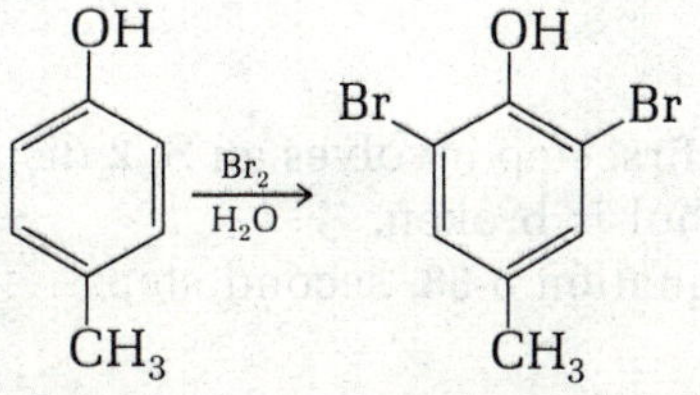

To show the solubility in water, write an equation analogous to equation 6-9.

6-17. *a*) The two transition states (see eq 6-17) would be:

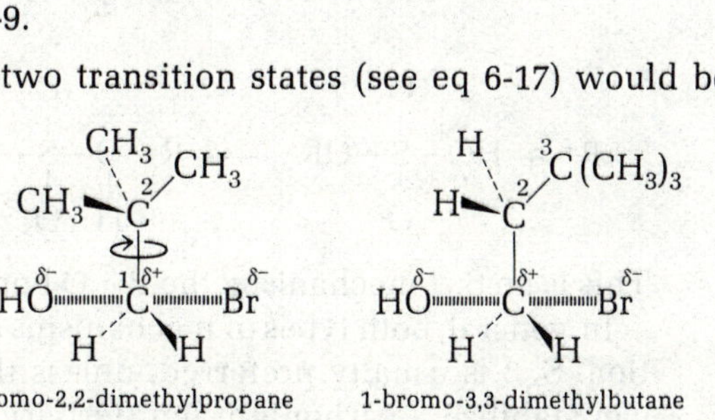

1-bromo-2,2-dimethylpropane   1-bromo-3,3-dimethylbutane

In the first case, approach of the OH⁻ from the "backside" of the carbon-bromine bond is severely hindered by methyl groups, no matter how the C1-C2 bond is rotated. On the other hand, in the second case the large group (C3 and its substituents) can be rotated out of the way of the attacking hydroxyl group.

b) An S_N2 process is impossible, since the "backside" of the carbon-chlorine bond is inaccessible (C4 is in the direct line of attack and prevents reaction by this path).

An S_N1 mechanism is also hindered, since the cage-like structure prevents carbons 2, 6 and 7 from becoming coplanar with carbon 1, as is required for a carbonium ion (eq 6-18a).

6-18. Equation 6-33, first step:

protonated
sulfuric acid

The first step involves an S_N2 displacement; the O—H bond of the alcohol is broken.

Equation 6-33, second step:

$$R—\ddot{O}H + H^+ \rightleftharpoons R—\overset{H}{\underset{..}{\overset{|}{O}}}{}^+—H \rightleftharpoons R^+ + H_2O$$

This is an S_N1 mechanism; the R—O bond of the alcohol is broken.

In general, both types of mechanisms are possible in ester formation. S_N2 is usually preferred, unless the R group of the alcohol can stabilize a carbonium ion (tertiary, benzylic or allylic).

6-19. *a)* $CH_3CH_2CH_2CH_2OH \xrightarrow[\text{heat}]{H^+} CH_3CH_2CH=CH_2 \xrightarrow[\text{sec 3.4c}]{HBr}$

$$CH_3CH_2\underset{\underset{Br}{|}}{C}HCH_3$$

b) $CH_3CH_2CH=CH_2 \xrightarrow[\substack{\text{eq 6-11} \\ \text{and 6-12}}]{\substack{H_2O \\ H^+}} CH_3CH_2\underset{\underset{OH}{|}}{C}HCH_3 \xrightarrow[\text{eq 6-4}]{Na}$

$$CH_3CH_2\underset{\underset{O^-Na^+}{|}}{C}HCH_3$$

c) $(CH_3)_2CHCH_2OH \xrightarrow[\text{eq 6-31}]{PBr_3} (CH_3)_2CHCH_2Br$

d) $CH_3-\!\!\bigcirc\!\!-OH \xrightarrow[\text{eq 6-39}]{\substack{H_2 \\ Ni}}$ [cyclohexanol structure with CH_3, H, H, OH] $\xrightarrow[\text{sec 3.5a}]{\substack{H^+ \\ \text{heat}}}$ [methylcyclohexene structure with CH_3, H]

e) $(CH_3)_3COH \xrightarrow[\text{sec 3.5a}]{\substack{H^+ \\ \text{heat}}} CH_2=C(CH_3)_2 \xrightarrow[\text{eq 3-10}]{\substack{H_2 \\ Ni}} CH_3-CH(CH_3)_2$

f) [cyclohexanol structure H, OH] $\xrightarrow[\text{sec 3.5a}]{\substack{H^+ \\ \text{heat}}}$ [cyclohexene structure] $\xrightarrow[\text{eq 6-40}]{HOCl}$ [cyclohexane with HO, H and H, Cl] $\xrightarrow[\substack{\text{base} \\ \text{eq 6-40}}]{\text{dil.}}$ [cyclohexane with HO, H and H, OH]

g) [2-methylnaphthalene with CH_3] $\xrightarrow[\text{eq 5-8}]{H_2SO_4}$ [naphthalene with SO_3H, CH_3] $\xrightarrow[\substack{\text{fusion} \\ \text{eq 6-21, 6-22}}]{NaOH}$ [naphthalene with OH, CH_3]

h) $CH_3CH_2OH \xrightarrow[\text{eq 6-27}]{\text{HBr}} CH_3CH_2Br \xrightarrow[\text{eq 6-44}]{\text{Na}^+\text{SH}^-} CH_3CH_2SH$

i)

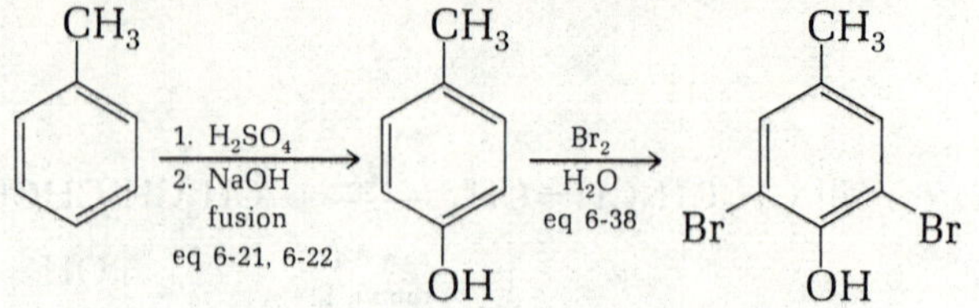

1. H_2SO_4
2. NaOH
   fusion
   eq 6-21, 6-22

$\xrightarrow[\text{eq 6-38}]{\overset{Br_2}{\underset{H_2O}{}}}$

j)

1. $(BH_3)_2$
2. H_2O_2
3. OH^-

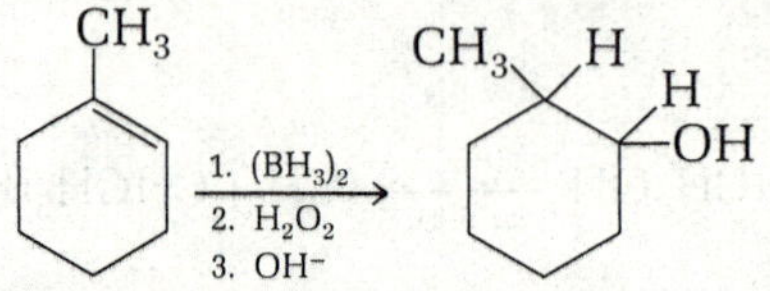

eq 6-13 and 6-14

6-20. This is an S_N2 displacement of halide by hydroxide, as in equation 6-17.

CHAPTER SEVEN

ETHERS

7-1. a) $CH_3CH_2CH_2OCH_2CH_2CH_3$

f) $HOCH_2CH_2OCH_2CH_2OH$

b) $CH_3OC(CH_3)_3$

g) $CH_3OCH_2CH_2OCH_3$

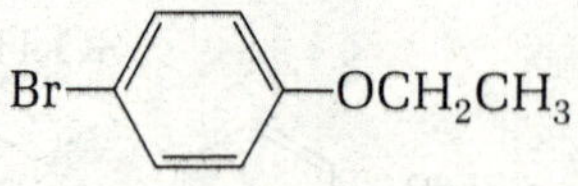

c) $CH_3CH_2CHCH_2CH_2CH_3$
   $\quad\quad\quad | $
   $\quad\quad OCH_3$

h) $(CH_3)_2CHSCH(CH_3)_2$

i) $ClCH_2OCH_2Cl$

d) $CH_2{=}CHCH_2OCH_2CH{=}CH_2$

j)

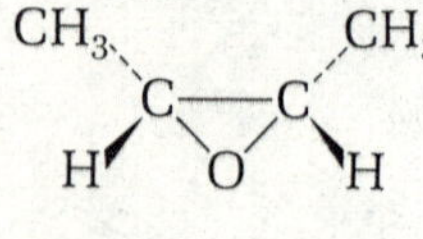

e) $Br{-}\langle\text{ring}\rangle{-}OCH_2CH_3$

7-2. a) isopropyl ether

f) phenyl *t*-butyl ether

b) methyl isobutyl ether

g) 2-ethoxypentane

c) propylene oxide

h) epichlorhydrin

d) *p*-bromoanisole (or *p*-bromo-phenyl methyl ether)

i) methyl *n*-propyl sulfide

j) methoxycyclopentane (or methyl cyclopentyl ether)

e) 2-methoxyethanol

7-3. In addition to the "crossed" ether, one would also obtain the simple ethers (*i.e.*, ethyl and *n*-propyl):

$$2 \ CH_3CH_2OH \xrightarrow{\ H^+ \ } CH_3CH_2OCH_2CH_3 + H_2O$$

$$2 \ CH_3CH_2CH_2OH \xrightarrow{\ H^+ \ } CH_3CH_2CH_2OCH_2CH_2CH_3 + H_2O$$

as well as

$$CH_3CH_2OH + CH_3CH_2CH_2OH \xrightarrow{\ H^+ \ } CH_3CH_2OCH_2CH_2CH_3 + H_2O$$

Review section 7.2a.

7-4. In the Williamson synthesis (eq 7-7), one R group of the ether comes from an alkyl halide, the other from an alkoxide. If the R groups differ, two combinations of reagents are possible:

a) $CH_3O^-Na^+ + CH_3CH_2Br \longrightarrow CH_3OCH_2CH_3 + Na^+Br^-$

$ CH_3CH_2O^-Na^+ + CH_3Br \longrightarrow CH_3OCH_2CH_3 + Na^+Br^-$

b) $CH_3O^-Na^+ + CH_3\underset{\underset{\displaystyle Br}{|}}{C}HCH_2CH_3 \longrightarrow$

$$CH_3\underset{\underset{\displaystyle OCH_3}{|}}{C}HCH_2CH_3 + Na^+Br^-$$

$$CH_3\underset{\underset{\displaystyle O^-Na^+}{|}}{C}HCH_2CH_3 + CH_3Br \longrightarrow CH_3\underset{\underset{\displaystyle OCH_3}{|}}{C}HCH_2CH_3 + Na^+Br^-$$

7-5. ⟨benzene ring⟩$-O^-Na^+ + CH_3CH_2CH_2CH_2Br \longrightarrow$

⟨benzene ring⟩$-OCH_2CH_2CH_2CH_3 + Na^+Br^-$

This reaction involves an S_N2 displacement of bromide ion from *n*-butyl bromide by the nucleophile, phenoxide ion.

If the converse pair of reagents were chosen—that is, sodium *n*-butoxide + bromobenzene—

$$CH_3CH_2CH_2CH_2O^-Na^+ + \text{⟨benzene ring⟩}-Br \longrightarrow \text{no reaction}$$

the nucleophile butoxide ion would have to displace bromide ion from bromobenzene. But S_N2 displacements of aryl bromides cannot occur, at least in the usual way, because "backside" approach to the carbon bearing the halogen is blocked by the aromatic ring. But even if the "backside" were not blocked, displacement of halogen attached to an sp^2 (rather than sp^3) carbon is difficult, due to resonance structures such as those shown in section 5.4b which place a + charge on the halogen, thus making it less likely to leave as halide ion. For a further discussion, look ahead to section 8.5.

7-6. a) $2\,CH_3CH_2OH \xrightarrow[140°]{H_2SO_4} CH_3CH_2OCH_2CH_3 + H_2O$ (eq 7-4)

b) no reaction; ethers (except for epoxides) are inert toward base.

c) $CH_3OCH_2CH_2CH_3 + 2\,HBr \longrightarrow CH_3Br + CH_3CH_2CH_2Br + H_2O$ (eq 7-13)

d) no reaction; ethers can be distinguished from alcohols by their inertness toward sodium metal.

e) $(CH_3)_3CO^-K^+ + CH_3CH_2I \longrightarrow (CH_3)_3COCH_2CH_3 + K^+I^-$ (eq 7-7)

f) $CH_3CH_2\overset{..}{\underset{..}{O}}CH_2CH_3 + H_2SO_4 \xrightarrow{cold} CH_3CH_2\overset{\overset{+}{..}}{\underset{\underset{H}{|}}{O}}CH_2CH_3 + HSO_4^-$

The ether acts as a base (eq 7-11) and dissolves in the strong acid.

7-7. The reaction is an example of an S_N2 displacement reaction.

This result gives some experimental evidence for the "backside" attack of nucleophiles in S_N2 displacements (see sec 6.5c).

Since methoxide ion is a base, a competing reaction is the elimination of HBr, to form an alkene (eq 3-31).

7-8. The mechanism by which the observed product (*t*-butyl methyl ether is formed is given in equation 7-6. Let us consider the possible paths to methyl ether or *t*-butyl ether.

For methyl ether:

$$CH_3\overset{..}{\underset{..}{O}}H + H^+ \rightleftharpoons [CH_3\overset{..}{\underset{\underset{H}{|}}{O}}H]^+$$

Protonated methanol will not readily lose H_2O, since the methyl carbonium ion CH_3^+ which would be formed is a primary carbonium ion. Therefore the only way methyl ether would be formed is by an S_N2 displacement mechanism:

$$CH_3\overset{..}{\underset{..}{O}}H + CH_3\overset{+}{O}H_2 \longrightarrow CH_3OCH_3 + H_2O$$

This reaction is possible, but usually requires a higher temperature than reactions which occur by the S_N1 path.

For t-butyl ether:

Here the S_N1 path is possible, except that the second step is

$$(CH_3)_3C^+ + (CH_3)_3COH \longrightarrow (CH_3)_3COC(CH_3)_3 + H^+$$

This would be much slower than the *observed* reaction

$$(CH_3)_3C^+ + CH_3OH \longrightarrow (CH_3)_3COCH_3 + H^+$$

because of steric hindrance in bringing two large *t*-butyl groups close to one another. This is readily seen if the formula for *t*-butyl ether is written out (or better still, if a model is constructed):

$$
\begin{array}{ccc}
 & O & \\
CH_3-C & & C-CH_3 \\
CH_3\;CH_3 & & CH_3\;CH_3
\end{array}
$$

7-9. Review section 7.5 before answering this question.

a) $\underset{\displaystyle O}{CH_2-CH_2} + HBr \longrightarrow HOCH_2CH_2Br$

b) $\underset{\displaystyle O}{CH_2-CH_2} + HO-\!\!\bigcirc \xrightarrow{H^+} HOCH_2CH_2O-\!\!\bigcirc$

c) $\underset{\displaystyle O}{CH_2-CH_2} + CH_3OH \xrightarrow{H^+} HOCH_2CH_2OCH_3$

If excess methanol is used, further reaction is possible, as in equations 7-3 and 7-4.

$$HOCH_2CH_2OCH_3 + HOCH_3 \xrightarrow{H^+} CH_3OCH_2CH_2OCH_3 + H_2O$$

d) The first step is as shown in equation 7-21:

$$\underset{\displaystyle O}{CH_2-CH_2} + NH_3 \longrightarrow HOCH_2CH_2NH_2$$

Ethanolamine can then function as the nucleophile in reacting with additional ethylene oxide:

$$\text{CH}_2\!\!-\!\!\text{CH}_2 + \text{H}_2\text{NCH}_2\text{CH}_2\text{OH} \longrightarrow \text{HOCH}_2\text{CH}_2\text{NHCH}_2\text{CH}_2\text{OH}$$

diethanolamine

Repetition gives the final product:

$$\text{CH}_2\!\!-\!\!\text{CH}_2 + \text{HN(CH}_2\text{CH}_2\text{OH})_2 \longrightarrow$$

$$\text{HOCH}_2\text{CH}_2\text{N(CH}_2\text{CH}_2\text{OH})_2 \text{ or } \text{N(CH}_2\text{CH}_2\text{OH})_3$$

triethanolamine

Certain salts of triethanolamine are used as emulsifying agents in agricultural sprays.

7-10. This question is based on equation 7-13. Since the product is a dibromide, the original ether must have been cyclic.

$$\xrightarrow{\text{2HBr}} \text{BrCH}_2\text{CH}_2\text{CH}_2\text{CH}_2\text{Br} + \text{H}_2\text{O}$$

tetrahydrofuran

7-11. *a*) Add a little of each compound, in separate test-tubes, to concentrated sulfuric acid. The ether will dissolve (eq 7-11) whereas the hydrocarbon, being inert and less dense than sulfuric acid, will simply float on top.

b) Add a little bromine in carbon tetrachloride to each ether. The allyl phenyl ether, being unsaturated, will quickly decolorize the bromine, whereas the ethyl phenyl ether will not.

$$\text{CH}_2\!\!=\!\!\text{CHCH}_2\text{OC}_6\text{H}_5 + \text{Br}_2 \xrightarrow[\text{eq 3-5}]{} \underset{\underset{\text{Br}}{|}}{\text{CH}_2}\!\!-\!\!\underset{\underset{\text{Br}}{|}}{\text{CHCH}_2}\text{OC}_6\text{H}_5$$

$$\text{CH}_3\text{CH}_2\text{OC}_6\text{H}_5 + \text{Br}_2 \longrightarrow \text{no reaction}$$

c) Add a small piece of sodium to each compound; the alcohol will liberate a gas (hydrogen), whereas no gas bubbles will be apparent in the ether.

$$2\,\underset{\underset{\text{OH}}{|}}{\text{CH}_3\text{CHCH}_2\text{CH}_3} + 2\,\text{Na} \xrightarrow[\text{eq 6-4}]{} 2\,\underset{\underset{\text{O}^-\text{Na}^+}{|}}{\text{CH}_3\text{CHCH}_2\text{CH}_3} + \text{H}_2$$

$$\text{CH}_3\text{OCH}_2\text{CH}_2\text{CH}_3 + \text{Na} \longrightarrow \text{no reaction}$$

(NOTE that the alcohol and ether are isomers.)

d) Add each compound to a little 10% aqueous sodium hydroxide. The phenol will dissolve, whereas the ether is inert toward base.

$$\text{\Large⬡}-\text{OH} + \text{Na}^+\text{OH}^- \longrightarrow \text{\Large⬡}-\text{O}^-\text{Na}^+ + \text{H}_2\text{O} \qquad \text{(eq 6-9)}$$

$$\text{\Large⬡}-\text{OCH}_3 + \text{Na}^+\text{OH}^- \longrightarrow \text{no reaction}$$

7-12. a) $2\ \text{CH}_3\text{CH}_2\text{CH}_2\text{CH}_2\text{OH} \xrightarrow[\text{heat}]{\text{H}_2\text{SO}_4}$

$$\text{CH}_3\text{CH}_2\text{CH}_2\text{CH}_2\text{OCH}_2\text{CH}_2\text{CH}_2\text{CH}_3 + \text{H}_2\text{O} \quad \text{(eq 7-3 and 7-4)}$$

Though the answers in the remaining parts to this question are given with the reactions in sequence, it is often a good idea to work backward. Write down the last step, the one which gives the desired product. Then see how you can get the necessary reagents for that step from the given starting materials.

b) $\text{\Large⬡} \xrightarrow[\substack{\text{2. NaOH}\\\text{fusion}\\\text{(sec 6.6a)}}]{\text{1. H}_2\text{SO}_4} \overset{\text{OH}}{\text{\Large⬡}} \xrightarrow[\text{(eq 6-9)}]{\text{Na}^+\text{OH}^-} \overset{\text{O}^-\text{Na}^+}{\text{\Large⬡}}$

$$\text{CH}_2{=}\text{CH}_2 + \text{HBr} \longrightarrow \text{CH}_3\text{CH}_2\text{Br} \qquad \text{(eq 3-11)}$$

$$\text{\Large⬡}-\text{O}^-\text{Na}^+ + \text{CH}_3\text{CH}_2\text{Br} \longrightarrow$$

$$\text{\Large⬡}-\text{OCH}_2\text{CH}_3 + \text{Na}^+\text{Br}^- \qquad \text{(eq 7-7)}$$

c) $\text{CH}_3\text{CH}_2\text{CH}_2\text{OCH}_2\text{CH}_2\text{CH}_3 + 2\ \text{HBr} \xrightarrow{\text{heat}}$

$$2\ \text{CH}_3\text{CH}_2\text{CH}_2\text{Br} + \text{H}_2\text{O} \qquad \text{(eq 7-13)}$$

$$\text{CH}_3\text{CH}_2\text{CH}_2\text{Br} \xrightarrow[\text{alcohol}]{\text{KOH in}} \text{CH}_3\text{CH}{=}\text{CH}_2 \qquad \text{(eq 3-31)}$$

d) $\text{CH}_2{=}\text{CH}_2 \xrightarrow[\text{H}^+]{\text{H}_2\text{O}} \text{CH}_3\text{CH}_2\text{OH} \qquad \text{(eq 6-11 and 6-12)}$

$$\text{CH}_2{=}\text{CH}_2 \xrightarrow[\text{2. OH}^-]{\text{1. HOCl}} \underset{\text{O}}{\text{CH}_2{-}\text{CH}_2} \qquad \text{(eq 7-17)}$$

$$\underset{\text{O}}{\text{CH}_2{-}\text{CH}_2} + \text{CH}_3\text{CH}_2\text{OH} \xrightarrow{\text{H}^+}$$

$$\text{CH}_3\text{CH}_2\text{OCH}_2\text{CH}_2\text{OH} \qquad \text{(eq 7-21)}$$

69

e)

$$\underset{\text{(eq 6-39)}}{\overset{\text{H}_2}{\xrightarrow{\quad\text{Ni}\quad}}} \qquad \underset{\text{(eq 6-4)}}{\overset{\text{Na}}{\xrightarrow{\qquad}}}$$

$$3\,CH_3CH_2OH + PBr_3 \xrightarrow[\text{eq 6-31}]{} 3\,CH_3CH_2Br + H_3PO_3$$

$$\text{(cyclohexyl-O}^-Na^+) + CH_3CH_2Br \xrightarrow[\text{eq 7-7}]{}$$

$$\text{(cyclohexyl-OCH}_2CH_3) + Na^+Br^-$$

f) $CH_3CH{=}CH_2 \xrightarrow[\text{heat}]{\text{Cl}_2} \underset{\displaystyle Cl}{CH_2CH{=}CH_2} \xrightarrow[\text{OH}^-]{\text{H}_2\text{O}}$
(eq 3-27a)   (eq 6-16)

$$HOCH_2CH{=}CH_2$$

$$2\,CH_2{=}CHCH_2OH \xrightarrow[\text{heat}]{\text{H}^+} CH_2{=}CHCH_2OCH_2CH{=}CH_2$$
(eq 7-2)

g) $CH_3CH_2Br + Na^+SH^- \xrightarrow[\text{(eq 6-44)}]{} CH_3CH_2SH + Na^+Br^-$

$$2\,CH_3CH_2SH + 2\,Na \xrightarrow[\text{(analogous to eq 6-4)}]{} 2\,CH_3CH_2S^-Na^+ + H_2$$

$$CH_3CH_2S^-Na^+ + CH_3CH_2Br \xrightarrow[\text{(eq 7-24)}]{}$$

$$CH_3CH_2SCH_2CH_3 + Na^+Br^-$$

h) $CH_2{=}CHCH_3 + HOCl \longrightarrow \underset{\displaystyle Cl \quad\ OH}{CH_2{-}CHCH_3}$ (eq 3-17)

$$\underset{\displaystyle Cl \quad\ OH}{CH_2{-}CHCH_3} \xrightarrow{\text{OH}^-} \underset{\displaystyle O}{CH_2{-}CHCH_3}$$ (eq 7-17)

7-13. Chlorine, in water, is a source of hypochlorous acid:

$$Cl_2 + H_2O \rightleftharpoons HCl + HOCl$$

$$CH_2{=}CH{-}CH_2Cl + \overset{\delta^-}{H}O{-}\overset{\delta^+}{Cl} \longrightarrow \underset{\displaystyle Cl}{CH_2{-}\overset{+}{CH}{-}CH_2Cl} + OH^-$$

$$\underset{\displaystyle Cl \quad\ OH}{CH_2{-}CH{-}CH_2Cl}$$

The first step follows Markownikoff's rule (eq 3-17).

In the second step, the base removes a proton from the alcohol function; the resulting alkoxide undergoes an intramolecular S_N2 displacement of chloride ion:

$$ClCH_2CHCH_2Cl + {}^-OH \rightleftharpoons ClCH_2CHCH_2Cl + H_2O$$
$$\qquad\quad | \qquad\qquad\qquad\qquad\qquad |$$
$$\qquad\quad OH \qquad\qquad\qquad\qquad\quad O^-$$

$$Cl-CH_2-CH-CH_2Cl \xrightarrow{-Cl^-} CH_2-CH-CH_2Cl$$
$$\qquad\qquad | \qquad\qquad\qquad\qquad\quad \diagdown O \diagup$$
$$\qquad\qquad O^-$$

epichlorhydrin

7-14. The reactions with alcohols are usually catalyzed by acid:

$$CH_2-CH_2 + H^+ \rightleftharpoons CH_2-CH_2$$
$$\diagdown \overset{..}{\underset{..}{O}} \diagup \qquad\qquad\qquad \diagdown \overset{+}{\underset{|}{O}} \diagup$$
$$\qquad\qquad\qquad\qquad\qquad\qquad H$$

Nucleophilic attack opens the ring:

$$CH_2-CH_2 + H\overset{..}{O}R \longrightarrow CH_2-CH_2-\overset{+}{O}-R$$
$$\diagdown \underset{+}{\overset{..}{O}} \diagup \qquad\qquad\qquad\qquad | \qquad\qquad\quad |$$
$$\qquad | \qquad\qquad\qquad\qquad\qquad OH \qquad\quad\ H$$
$$\quad H$$

$$\Big\Updownarrow -H^+$$

$$HOCH_2CH_2OR$$

These equations represent the mechanism for the first two parts of equation 7-21, where R = CH_3- or $HOCH_2CH_2-$.

Ammonia or amines attack the ethylene oxide directly:

$$CH_2-CH_2 + :NH_3 \longrightarrow CH_2-CH_2-\overset{+}{N}H_3$$
$$\diagdown O \diagup \qquad\qquad\qquad\qquad | $$
$$\qquad\qquad\qquad\qquad\qquad\qquad O^-$$

$$CH_2CH_2\overset{+}{N}H_3 \underset{\text{a proton}}{\overset{\text{remove and add}}{\rightleftharpoons}} CH_2CH_2NH_2$$
$$| \qquad\qquad\qquad\qquad\qquad\qquad |$$
$$O^- \qquad\qquad\qquad\qquad\qquad\quad OH$$

7-15. This reaction is a nucleophilic displacement on an ethylene oxide:

71

The alkoxide in the middle of the formula picks up a proton from the solvent (water or alcohol) to become a hydroxyl group. The phenoxide ion at the right end of the molecule is ready to act as a nucleophile and repeat the process.

The chain can be terminated by ethylene oxide units if, in the last step, a molecule of bisphenol A is displaced intramolecularly:

7-16. The starting compound has three functional groups.

a) The double bond is oxidized (Baeyer's test), the product being

$$\begin{array}{cccc} CH_2 & CH & CH & CH_2 \\ | & | & | & | \\ OH & OH & OCH_3 & OH \end{array}$$ or a further oxidation product of this.

b) The alcohol function reacts:

$$2\ CH_2{=}CH{-}\underset{\underset{OCH_3}{|}}{CH}{-}CH_2OH + 2\ K \longrightarrow$$

$$2\ CH_2{=}CH{-}\underset{\underset{OCH_3}{|}}{CH}{-}CH_2O^-K^+ + H_2{\uparrow}$$

c) None of the functional groups reacts with alkali.

d) Excess HBr might react with ALL THREE functional groups:

$$CH_2{=}CH{-}\underset{\underset{OCH_3}{|}}{CH}{-}CH_2OH \xrightarrow{4\ HBr}$$

addition   cleavage   displacement

$$CH_3{-}\underset{\underset{Br}{|}}{CH}{-}\underset{\underset{Br}{|}}{CH}{-}CH_2Br + CH_3Br + 2\ H_2O$$

e) Bromine adds to the double bond; the product is

$$\begin{array}{cccc} CH_2 & CH & CH & CH_2OH \\ | & | & | & \\ Br & Br & OCH_3 & \end{array}$$

7-17. Since the product has only two carbons, and the starting material has four carbons, two groups of two carbons must be separated by an ether oxygen.

$$-C-C-O-C-C-$$

The remaining two oxygens ($C_4H_{10}O_3$) must be at the ends of the chain. The desired structure is

$$HO-CH_2-CH_2-O-CH_2-CH_2-OH$$

and the equation for the reaction with HBr is

$$HOCH_2CH_2OCH_2CH_2OH + 4\,HBr \longrightarrow 2\,BrCH_2CH_2Br + 3\,H_2O$$

(Compare with equations 6-27 and 7-13.)

7-18. The relatively low hydrogen/carbon ratio and the fact that it absorbs three moles of H_2 suggest that **A** is aromatic. The molecular formula of **B** must be $C_8H_{16}O_2$ (**A** + 3 H_2). **B** is therefore a substituted cyclohexane.

The reaction of **B** with excess hot HBr suggests that **B** is an ether. The loss of two carbons (**B** $\longrightarrow$ **C**) and methyl bromide suggest that **B** is a dimethoxycyclohexane, and that **C** is a dibromo-cyclohexane:

$C_8H_{16}O_2$
B

$C_6H_{10}Br_2$
C

This implies that **A** is a dimethoxybenzene:

$C_8H_{10}O_2$
A

The only question which remains is the orientation of the methoxyl groups. Since **A** gives only one monobromo derivative, the methoxyls must be *para*.

The equations are given on the following page.

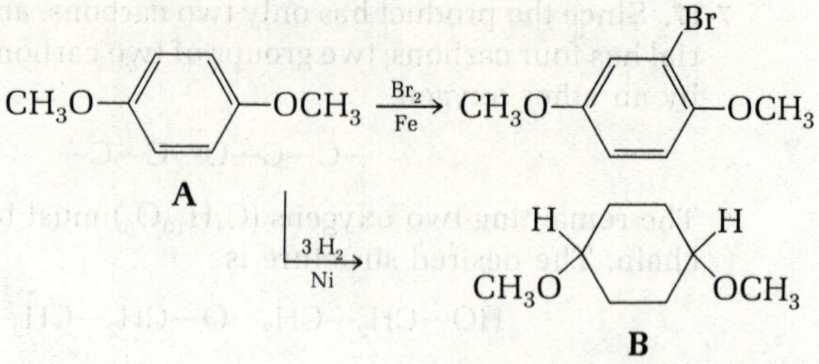

A

B

(a mixture of *cis* and *trans* isomers)

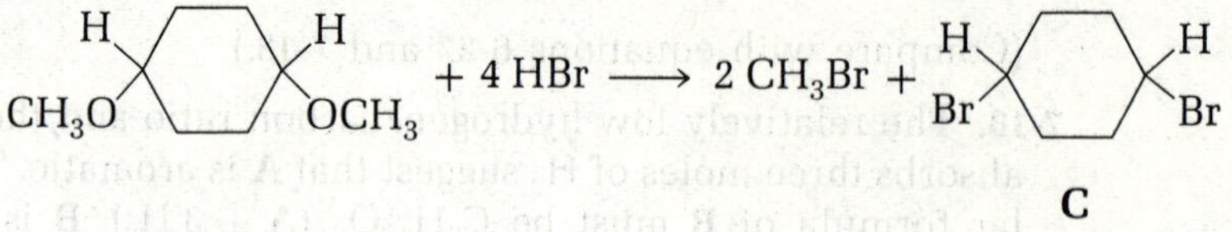

C

(a mixture of *cis* and *trans* isomers)

CHAPTER EIGHT

ORGANIC HALOGEN COMPOUNDS

8-1. *a)*

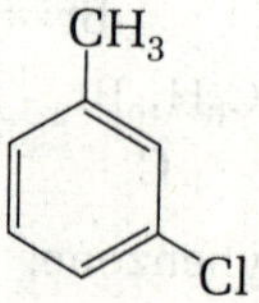

b) CH₃CH—CHCH₃
           | |
          Br   Br

c) CHBr₃

d)

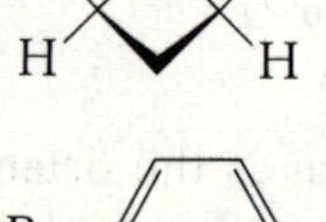

e)

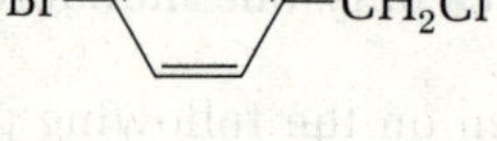

f) CH₂=CHCH₂I

g) CFCl₃

h)

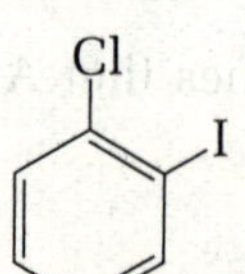

i) CH₂=CHBr

j) CH₃CH₂CH₂CH₂F

8-2. *a*) 2,2-dimethyl-1-bromopropane
(also called neopentyl bromide)

b) *p*-bromochlorobenzene
(alphabetic order of substituents

c) 4-bromo-3-methyl-1-chloropentane
(if the bromine is given preference, the
numbers will be higher).

d) 3-phenyl-1-chloropropane

e) 2,2-difluorobutane

f) 1,4-dichloronaphthalene

g) 1,4-dibromo-2-butyne

h) *n*-propylmagnesium chloride

i) *p*-chloroanisole

j) 2-bromopropene

8-3. Review section 2.7a

$$Br_2 \xrightarrow{\text{sunlight}} 2\,Br\cdot \quad \text{(initiation)}$$

chain
propagating
steps

Of the five different "kinds" of hydrogens in cumene, only the one
on the carbon α to the aromatic ring is abstracted, since the resulting radical is resonance-stabilized (sec 5.5).

8-4. See the answer to question 5-19.

8-5. *a*) $CH_3CHCH_3 + HCl \xrightarrow[\text{(conc)}]{ZnCl_2} CH_3CHCH_3 + H_2O \quad \text{(eq 6-27)}$
   $\overset{|}{OH} \qquad\qquad\qquad\qquad \overset{|}{Cl}$

b) $CH_3CHCH_3 \xrightarrow[\text{heat}]{H^+} CH_3CH{=}CH_2 \xrightarrow[\text{(eq 3-27a)}]{Cl_2} CH_2CH{=}CH_2$
   $\overset{|}{OH} \;\text{(sec 3.5a)} \qquad\qquad\qquad\qquad 400° \quad \overset{|}{Cl}$

c) $CH_3CH{=}CH_2 + Br_2 \xrightarrow{\text{(eq 3-2)}} CH_3CH{-}CH_2$
   (from *b*) $\qquad\qquad\qquad\qquad \overset{|}{Br}\;\; \overset{|}{Br}$

d) $2\,CH_3CHCH_3 + 2\,Na \longrightarrow 2\,CH_3CHCH_3 + H_2$ ⎫
      |OH |O⁻Na⁺

$CH_3CHCH_3 + CH_3I \longrightarrow CH_3CHCH_3 + Na^+I^-$ ⎬ sec 7.2b
    |O⁻Na⁺ |OCH₃ ⎭

e) $CH_3CH{=}CH_2 + HOCl \longrightarrow CH_3CHCH_2Cl$ (eq 3-17)
   (from b) |OH

f) CH_3CHCH_3 + ⬡ $\xrightarrow[\text{Friedel-Crafts Alkylation}]{\text{AlCl}}$ ⬡ + HCl
  |Cl |CH(CH₃)₂
  (from a)

8-6. Each of these reactions involves displacement of a halogen by a nucleophile; review section 8.2.

a) $CH_3CH_2CH_2CH_2Br + NaI \xrightarrow{\text{acetone}} CH_3CH_2CH_2CH_2I + NaBr$

b) $CH_3CHCH_2CH_3 + Na^{+-}OC_2H_5 \longrightarrow$
    |Cl

                             $CH_3CHCH_2CH_3 + Na^+Cl^-$
                                   |OC₂H₅

(the Williamson ether synthesis)

c) $(CH_3)_3CBr + H_2O \longrightarrow (CH_3)_3COH + H_2O$

The mechanism here is S_N1 (most of the other reactions in this section occur by an S_N2 mechanism).

d) ⬡—CH₂Br + NaCN $\longrightarrow$ ⬡—CH₂CN + Na⁺Br⁻

e) $CH_3CH_2CH_2I + Na^{+-}C{\equiv}CH \longrightarrow$

                           $CH_3CH_2CH_2C{\equiv}CH + Na^+I^-$

f) $CH_3CHCH_3 + NaSH \longrightarrow CH_3CHCH_3 + Na^+Cl^-$
   |Cl |SH

g) $CH_2{=}CHCH_2Cl + NaNH_2 \longrightarrow CH_2{=}CHCH_2NH_2 + Na^+Cl^-$

h) $CH_2CH_2CH_2CH_2 + 2\,NaCN \longrightarrow$
  |Br |Br

                       $CH_2CH_2CH_2CH_2 + 2\,Na^+Br^-$
                         |CN |CN

Displacement occurs at both possible positions.

i) ⬡(CH$_3$)(Br) + CH$_3$CH$_2$CH$_2$OH $\longrightarrow$

⬡(CH$_3$)(OCH$_2$CH$_2$CH$_3$) + HBr

The starting halide is tertiary, and the mechanism is S$_N$1.

j) No reaction; aromatic halides are not readily displaced by nucleophiles, unless the aromatic ring also contains electron-withdrawing substituents (sec 8.5).

8-7. *a)* CH$_3$CH$_2$CH$_2$CH$_3$ + Br$_2$ $\xrightarrow[\text{light}]{\text{heat} \atop \text{or}}$ CH$_3$CHCH$_2$CH$_3$ + HBr
|
Br

The product will be contaminated with 1-bromobutane, as well as with di-, tri-, etc. bromobutanes.

b) CH$_3$CHCH$_2$CH$_3$ + HBr $\xrightarrow{\text{ZnBr}_2}$ CH$_3$CHCH$_2$CH$_3$ (sec 6.7a)
| |
OH Br

One could also use PBr$_3$ as the reagent. This method gives a good yield and is unlikely to give isomers.

c) CH$_2$=CHCH$_2$CH$_3$ + HBr $\longrightarrow$ CH$_3$CHCH$_2$CH$_3$
|
Br

Addition follows Markownikoff's rule (sec 3.4c) and gives mainly the desired isomer.

d) CH$_3$CH=CHCH$_3$ + HBr $\longrightarrow$ CH$_3$CHCH$_2$CH$_3$
|
Br

The alkene is symmetric and can give only one addition product.

8-8. In each case, write the structural formula of the desired product, note the positions of the halogens, and select the desired alkene, alkyne, or diene accordingly.

a) CH$_2$=CHCH$_2$CH$_3$ + Br$_2$ $\longrightarrow$ CH$_2$—CHCH$_2$CH$_3$
| |
Br Br

b) HC≡CH $\xrightarrow{\text{HCl}}$ CH$_2$=CHCl $\xrightarrow{\text{HCl}}$ CH$_3$CHCl$_2$ (eq 4-4)

c) CH$_2$=CH—CH=CH$_2$ + 2 Br$_2$ $\longrightarrow$ CH$_2$CH—CH—CH$_2$
| | | |
Br Br Br Br

d) + HI $\longrightarrow$ —I

e) $CH_2{=}CH{-}CH{=}CH_2 + Br_2 \longrightarrow$

$$\underset{Br}{CH_2}{-}CH{=}CH{-}\underset{Br}{CH_2} \qquad \text{(eq 4-22)}$$

f) $HC{\equiv}C{-}CH_3 + 2\,Cl_2 \longrightarrow HC{-}\overset{\overset{\displaystyle Cl}{|}}{\underset{\underset{\displaystyle Cl}{|}}{C}}{-}CH_3 \qquad \text{(eq 4-2)}$

g) —CH=CH$_2$ + HBr $\longrightarrow$ —$\underset{Br}{CHCH_3}$ (sec 3.4c)

h) + 2 Br$_2$ $\longrightarrow$

8-9. The first step in the hydrolysis of any one of these halides is the ionization to a *t*-butyl cation:

$$(CH_3)_3C{-}X \xrightarrow{\;H_2O\;} (CH_3)_3C^+ + X^-$$
$$(X = Cl, \; Br \; or \; I)$$

The product-determining step involves the partition of this intermediate between two paths—reaction with water, or loss of a proton:

$$(CH_3)_3COH \xleftarrow{\;H_2O\;} (CH_3)_3C^+ \xrightarrow{\;-H^+\;} CH_2{=}C(CH_3)_2$$

Since the halide ion is, to a first approximation, not involved in these steps, this partition will occur in the same ratio regardless of which alkyl halide was being hydrolyzed.

8-10. Review section 8.3 before answering this question.

If only water is present as the nucleophile, the reaction follows an S_N1 path:

$$\underset{Br}{CH_2{=}CHCHCH_3} \xrightarrow{\;H_2O\;}$$

$$[CH_2{=}CH{-}\overset{+}{C}HCH_3 \longleftrightarrow \overset{+}{C}H_2{-}CH{=}CHCH_3]$$

allylic carbonium ion

The allylic carbonium ion can react with water at either of two sites, giving two different alcohols:

$$CH_2{=}CH{-}\underset{\underset{OH}{|}}{C}HCH_3 \quad \text{and} \quad CH_2CH{=}CHCH_3$$
$$\underset{OH}{|}$$

When the nucleophile is hydroxide ion (20% NaOH), the mechanism changes to S_N2; only the alcohol corresponding to the original halide is produced.

$$CH_2{=}CH\underset{\underset{Br}{|}}{C}HCH_3 + {}^-OH \longrightarrow CH_2{=}CH\underset{\underset{OH}{|}}{C}HCH_3 + Br^-$$

8-11. Review section 8.4 before answering this question.

Cyclohexyl chlorides only have the transoid coplanar geometry of the E2 transition state when the chlorine is in an axial position:

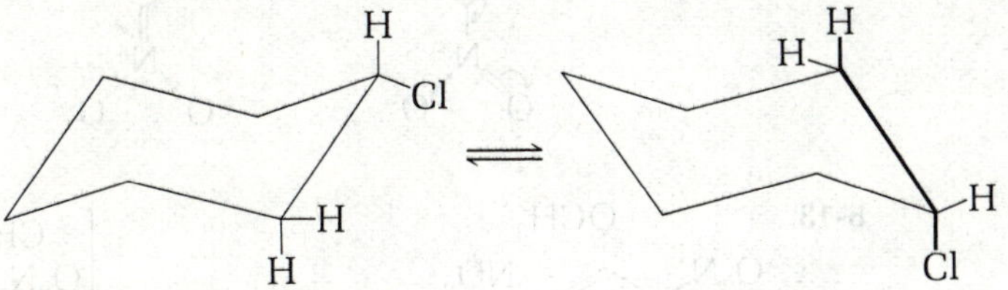

Consider menthyl chloride:

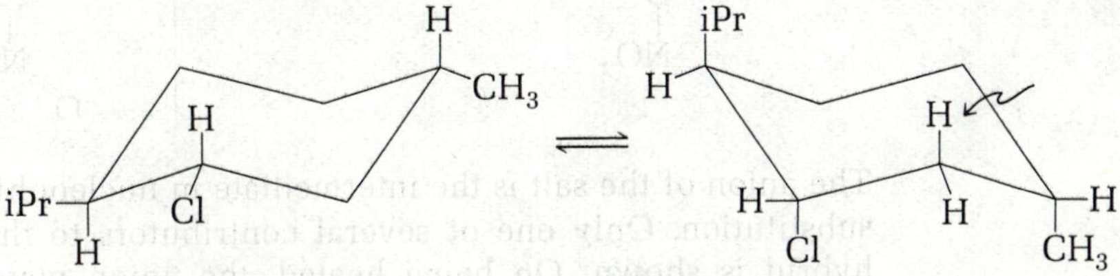

In the conformation at the right, with chlorine axial, the only hydrogen on an adjacent carbon suitably located for E2 elimination is shown by the arrow. Therefore the product is 2-menthene.

Consider neomenthyl chloride:

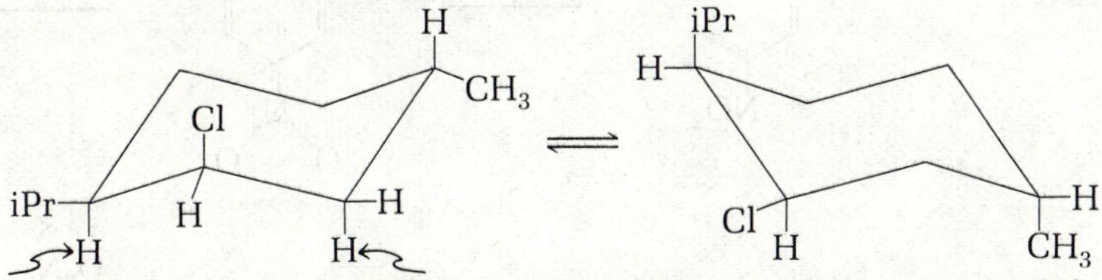

In this case, the conformation with chlorine axial is on the left. Two hydrogens (marked with arrows) have the suitable geometry for E2 elimination, and both are eliminated (75% 3-menthene, 25% 2-menthene).

8-12.

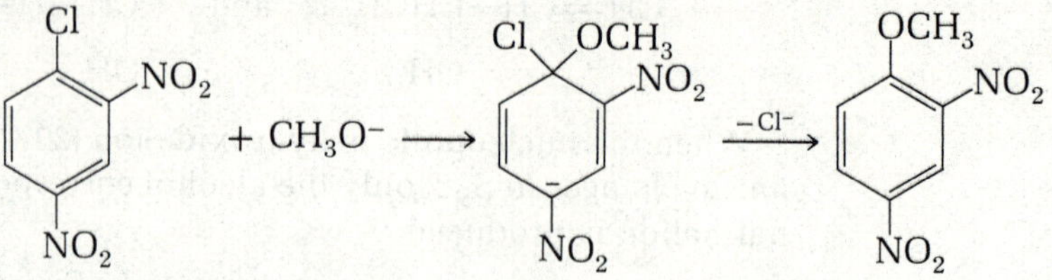

$$\text{+ CH}_3\text{O}^- \longrightarrow \quad \xrightarrow{-\text{Cl}^-}$$

The intermediate is stabilized by the nitro groups, through such resonance contributors as:

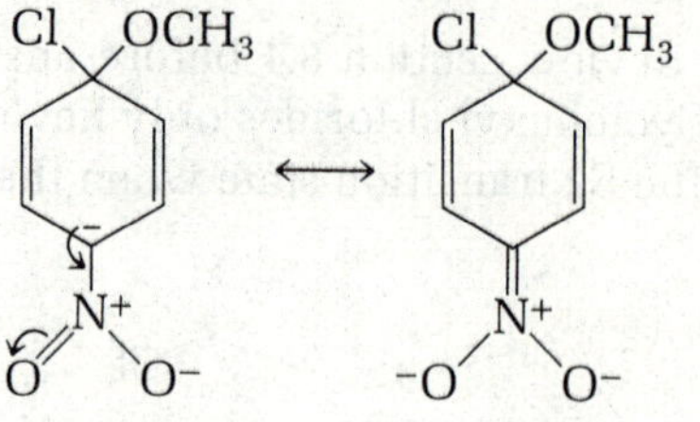

8-13.

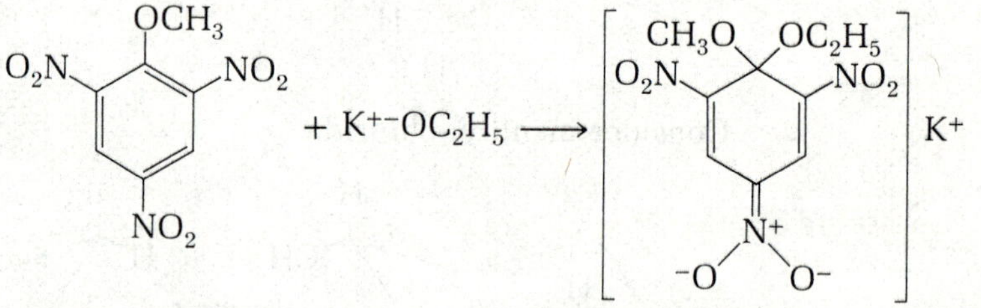

$$\text{+ K}^+{}^-\text{OC}_2\text{H}_5 \longrightarrow \qquad \text{K}^+$$

The anion of the salt is the intermediate in nucleophilic aromatic substitution. Only one of several contributors to the resonance hybrid is shown. On being heated, the anion may lose either methoxide ion or ethoxide ion.

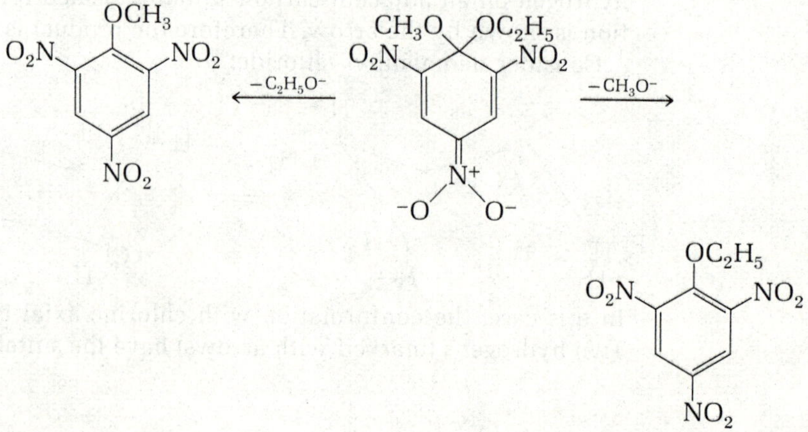

$$\xleftarrow{-\text{C}_2\text{H}_5\text{O}^-} \qquad \xrightarrow{-\text{CH}_3\text{O}^-}$$

This experiment provides support for the proposed mechanism for nucleophilic aromatic substitution.

8-14. The two intermediates are:

and

In each case, if the nitro group is to assist in delocalizing the negative charge it must be coplanar with the aromatic ring (just as with the C=C bond, the four atoms attached to a C=N bond must also lie in a single plane; see sec 3.2). This is difficult if there are large substituents (methyl groups) *ortho* to the nitro group. For this reason, 4-nitrobromobenzene is the more reactive of the two bromo compounds in nucleophilic displacement reactions.

8-15. *a)* $CH_3CHCH_2CH_3 \xrightarrow[\text{Ether}]{\text{Mg}} CH_3CHCH_2CH_3 \xrightarrow[\text{(eq 8-34)}]{H_2O,\ H^+}$

with Br below first, MgBr below second

$$CH_3CH_2CH_2CH_3 + Mg(OH)Br$$

b) $CH_2{=}CHCH_2Br \xrightarrow[\text{Ether}]{\text{Mg}} CH_2{=}CHCH_2MgBr \xrightarrow[\text{(eq 8-36)}]{\begin{smallmatrix}CH_2-CH_2\\ \diagdown O \diagup \end{smallmatrix}}$

$$CH_2{=}CHCH_2CH_2CH_2OMgBr \xrightarrow[H^+]{H_2O}$$

$$CH_2{=}CHCH_2CH_2CH_2OH$$

c) $CH_3CH_2CH_2OH \xrightarrow{HBr} CH_3CH_2CH_2Br \xrightarrow{Mg}$

$$CH_3CH_2CH_2MgBr \xrightarrow[\text{(eq 8-34)}]{D_2O} CH_3CH_2CH_2D + Mg(OD)Br$$

d)

$-CH_2CH_2OH \xrightarrow[\text{2. Mg, ether}]{\text{1. HBr}}$

$-CH_2CH_2MgBr \xrightarrow[\text{(eq 8-30)}]{\begin{smallmatrix}CH_2-CH_2\\ \diagdown O \diagup \end{smallmatrix}}$

$-CH_2CH_2CH_2CH_2OMgBr \xrightarrow[H^+]{H_2O}$

$-CH_2CH_2CH_2CH_2OH$

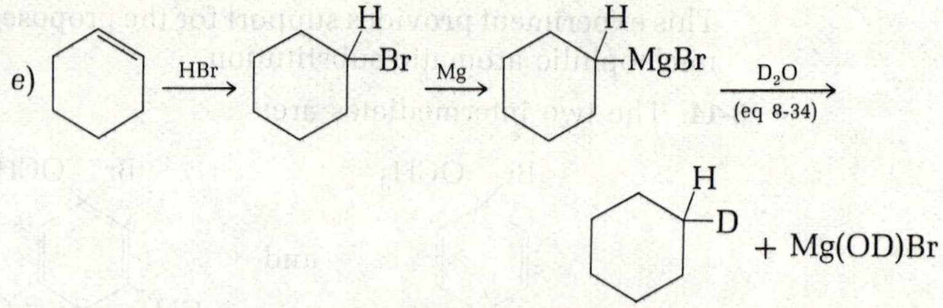

e)

$$\text{HBr} \quad \text{Mg} \quad \xrightarrow{D_2O} \text{(eq 8-34)}$$

$$+ \text{Mg(OD)Br}$$

8-16. Two reaction types are involved: addition to multiple bonds (sec 3.4a) and E2 eliminations (sec 8.3). There is nothing exceptional about the mechanism of the former. The two elimination steps are:

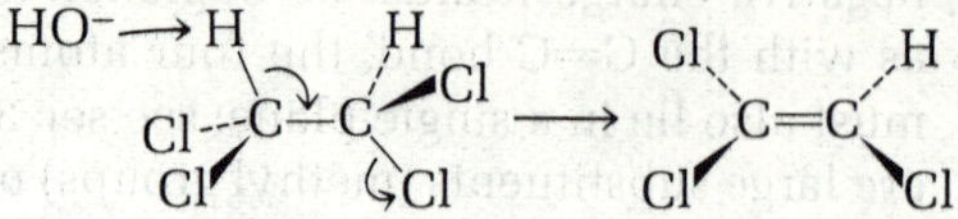

and

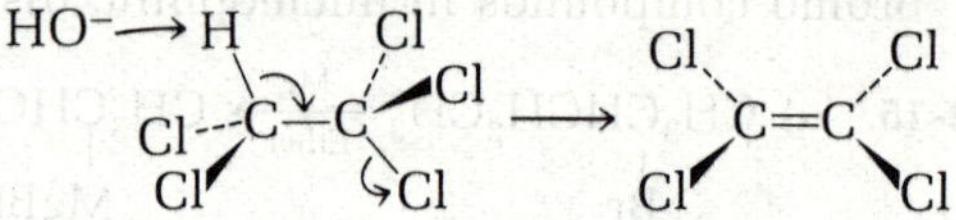

8-17. The reaction involves a divalent carbon intermediate, produced as follows:

$$\text{CHCl}_3 + {}^-\text{OH} \longrightarrow {}^-\text{CCl}_3 + \text{H}_2\text{O}$$

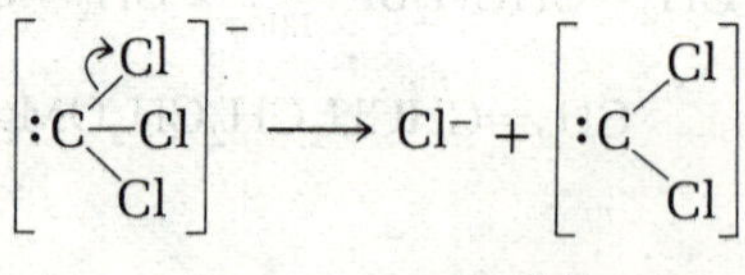

dichlorocarbene

The carbene, as it is called, is extremely reactive, and adds to double bonds:

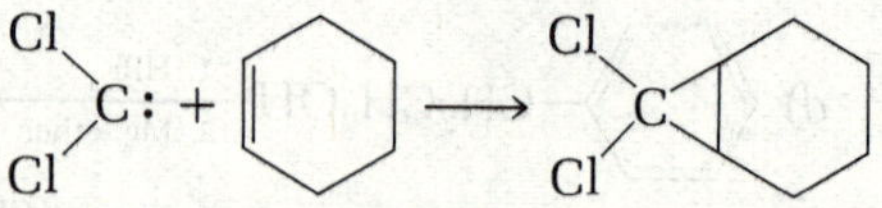

Carbenes form an important class of reactive organic intermediates (others are carbonium ions, carbanions, free radicals).

CHAPTER NINE

ALDEHYDES AND KETONES

9-1. *a)* 3-pentanone

b) hexanal

c) benzophenone
(also, diphenyl ketone)

d) p-bromobenzaldehyde

e) cyclopentanone

f) 2,2-dimethylpropanal

g) crotonaldehyde
(or 2-butenal)

h) 3-penten-2-one

i) bromoacetone
(or bromopropanone)

j) 3-butanol-2-one
(give the lower number to
the carbonyl group)

9-2. *a)* $\underset{\text{O}}{\overset{\text{O}}{\parallel}}$ $CH_3CCH_2CH_2CH_2CH_2CH_2CH_3$

b) $(CH_3)_2CHCH_2CH_2CHO$

c)

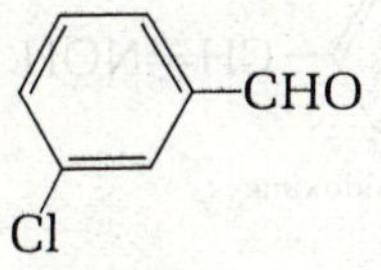

3-chlorobenzaldehyde structure: benzene ring with —CHO and Cl substituent

d)

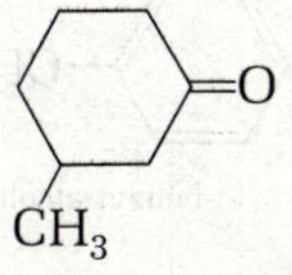

cyclohexanone ring with CH₃ substituent and =O

e) $CH_3CH=CHCHO$

f)

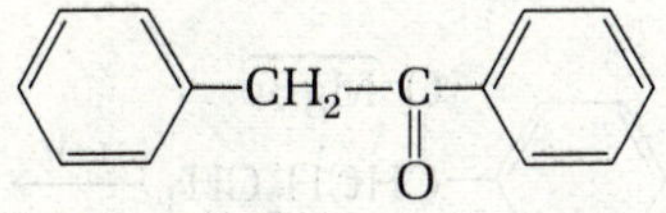

phenyl—CH_2—C(=O)—phenyl

g) CH_3—(benzene ring)—CHO

h) HO—(benzene ring)—$\underset{O}{\overset{O}{\parallel}}C$—(benzene ring)—$OH$

83

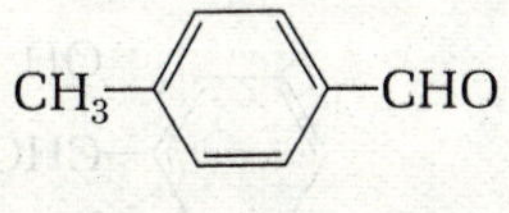

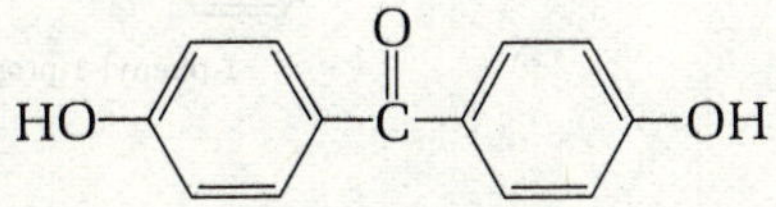

i) $CH_3(CH_2)_3CBr_2CHO$

j)

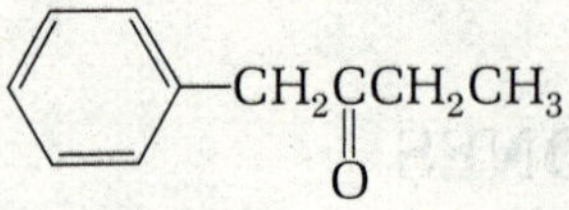

$\text{Ph—}CH_2CCH_2CH_3$
$\quad\quad\quad\overset{\|}{O}$

9-3. *a–d*) Sec 9.6c   *g–i*) Table 9-3
  e) Sec 9.6e   *j*) Sec 9.7
  f) Sec 9.4a

9-4. *a)* 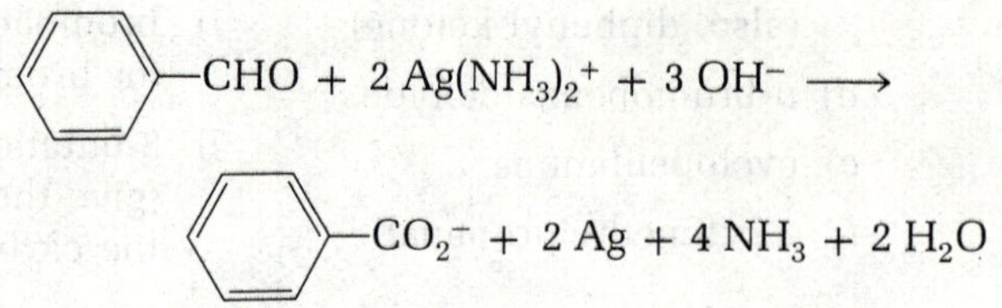 Ph—$CHO + 2\,Ag(NH_3)_2{}^+ + 3\,OH^- \longrightarrow$

  Ph—$CO_2{}^- + 2\,Ag + 4\,NH_3 + 2\,H_2O$ (eq 9-14)

  benzoate ion

b) 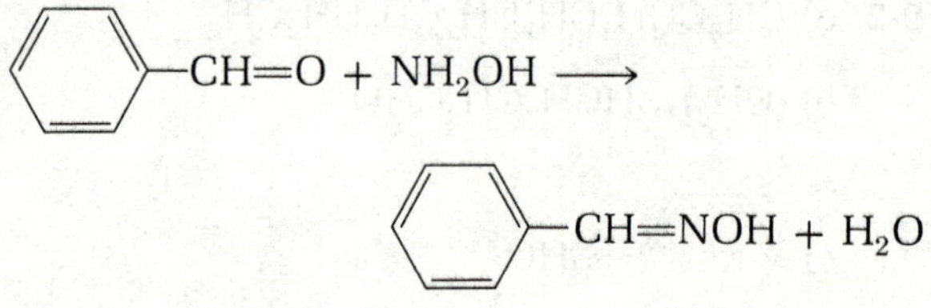 Ph—$CH{=}O + NH_2OH \longrightarrow$

  Ph—$CH{=}NOH + H_2O$ (eq 9-29)

  benzaldoxime

c) 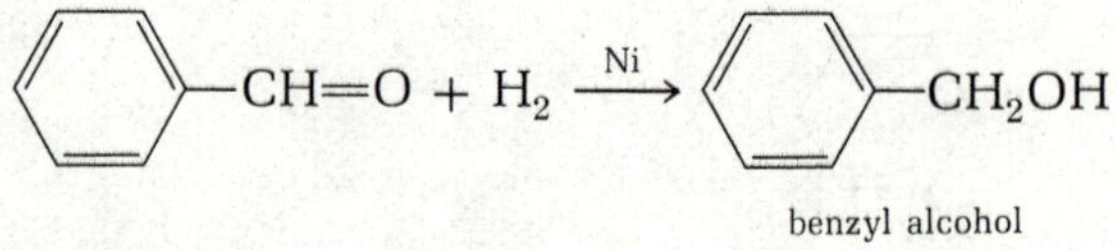 Ph—$CH{=}O + H_2 \xrightarrow{Ni}$ Ph—CH_2OH (eq 9-41)

  benzyl alcohol

d) 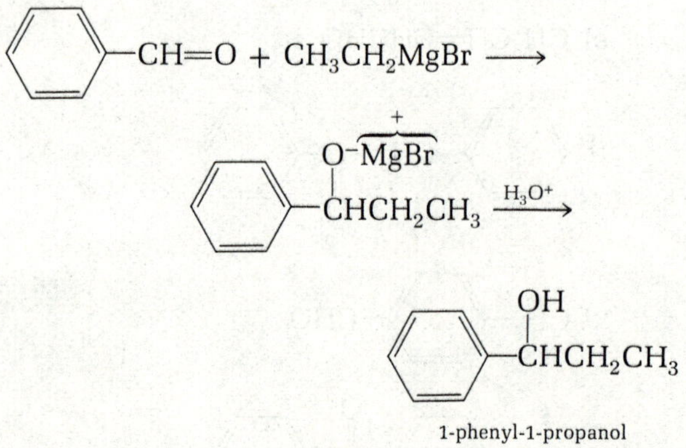 Ph—$CH{=}O + CH_3CH_2MgBr \longrightarrow$

  $\overset{+}{O^-MgBr}$
  Ph—$CHCH_2CH_3 \xrightarrow{H_3O^+}$

  OH
  Ph—$CHCH_2CH_3$ (eq 9-36)

  1-phenyl-1-propanol

e) 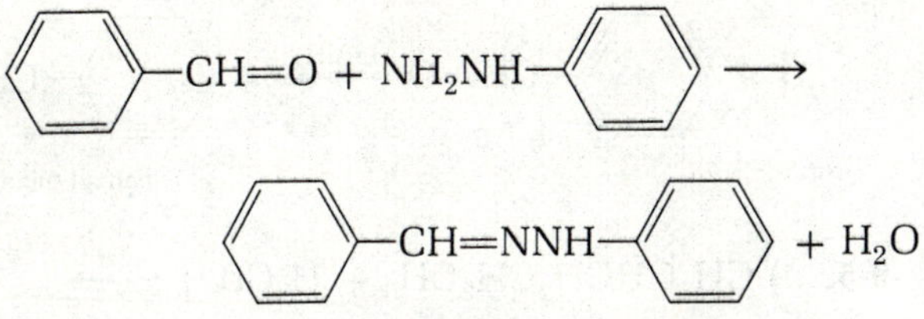 [benzene ring]—CH=O + NH₂NH—[benzene ring] ⟶

[benzene ring]—CH=NNH—[benzene ring] + H₂O   (eq 9-30)

benzaldehyde phenylhydrazone

f) 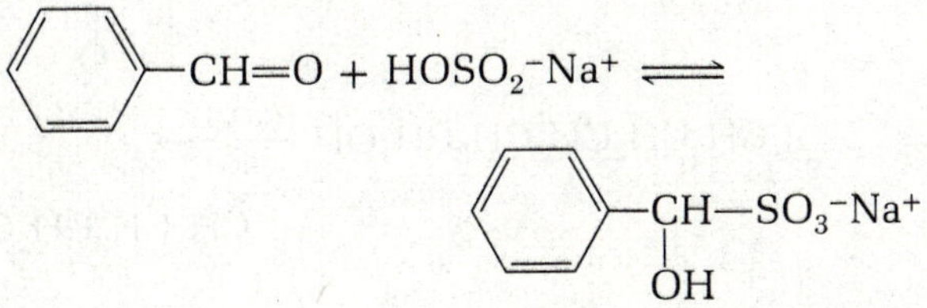 [benzene ring]—CH=O + HOSO₂⁻Na⁺ ⇌

[benzene ring]—CH—SO₃⁻Na⁺   (eq 9-40)
                  |
                  OH

benzaldehyde-bisulfite addition salt

g) 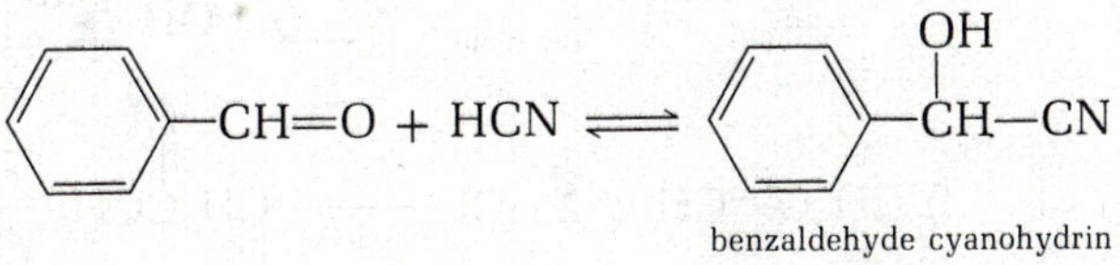 [benzene ring]—CH=O + HCN ⇌ [benzene ring]—CH—CN   (eq 9-31)
                                              |
                                              OH
                                              (OH above)

benzaldehyde cyanohydrin

h) 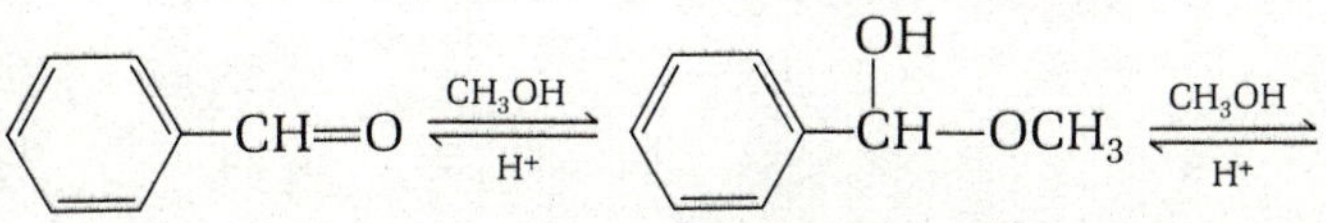 [benzene ring]—CH=O $\xrightarrow[\text{H}^+]{\text{CH}_3\text{OH}}$ [benzene ring]—CH—OCH₃ $\xrightarrow[\text{H}^+]{\text{CH}_3\text{OH}}$
                                                        |
                                                        OH

benzaldehyde
methyl hemiacetal

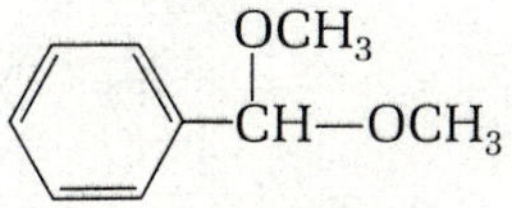

[benzene ring]—CH—OCH₃   (eq 9-25)
                |
                OCH₃

benzaldehyde
methyl acetal

i) 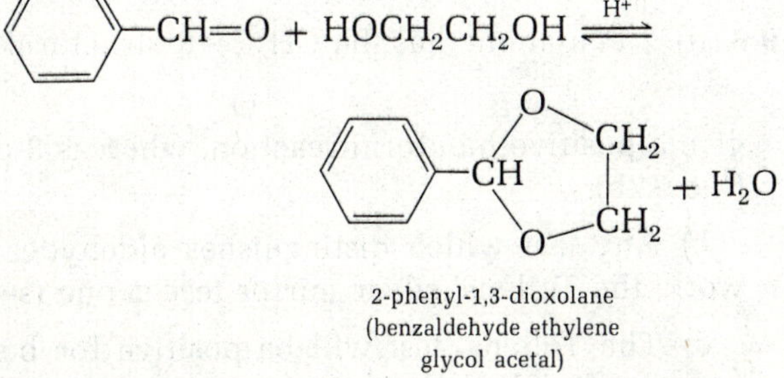 [benzene ring]—CH=O + HOCH₂CH₂OH $\xrightarrow{\text{H}^+}$ ⇌

[benzene ring]—CH⟨O—CH₂ / O—CH₂⟩ + H₂O   (eq 9-26)

2-phenyl-1,3-dioxolane
(benzaldehyde ethylene
glycol acetal)

j) 4 ⬡—CH=O $\xrightarrow{\text{LiAlH}_4}$ 4 ⬡—CH$_2$OH   (eq 9-43)

benzyl alcohol

9-5. *a)* $CH_3CHCH_2CH_2CH_3 + H_2CrO_4 \xrightarrow{\text{heat}}$
　　　　　|
　　　　　OH

$$CH_3CCH_2CH_2CH_3 \quad \text{(eq 9-4)}$$
　　　　‖
　　　　O

b) $CH_3CH_2CH_2CH_2CH_2OH \xrightarrow[\text{heat}]{\text{CrO}_3,\text{H}^+}$

$$CH_3CH_2CH_2CH_2CHO \quad \text{(eq 9-5)}$$

c) ⬡—OH $\xrightarrow[\text{(eq 6-39)}]{\underset{\text{Ni}}{3\text{H}_2}}$ (cyclohexane with H and OH) $\xrightarrow[\text{(eq 9-4)}]{\underset{\text{H}^+}{\text{CrO}_3}}$ (cyclohexanone)=O

d) $CH_3C{\equiv}CH + 2\,HCl \xrightarrow{\text{(eq 4-5)}}$ $CH_3\overset{\overset{\text{Cl}}{|}}{\underset{\underset{\text{Cl}}{|}}{C}}CH_3 \xrightarrow[\text{H}_2\text{O}]{\text{OH}^-}$

$$CH_3CCH_3 \quad \text{(eq 9-7)}$$
　　　　‖
　　　　O

e) ⬡—CH$_3$ $\xrightarrow[\text{(eq 5-8)}]{\underset{\text{FeCl}_3}{\text{Cl}_2}}$ Cl—⬡—CH$_3$ $\xrightarrow[\text{UV light}]{2\text{Cl}_2}$

Cl—⬡—CHCl$_2$ $\xrightarrow{\text{OH}^-}$ Cl—⬡—CHO   (eq 9-6)

f) $CH_2{=}CH(CH_2)_4CH_3 + CO + H_2 \xrightarrow{[\text{Co(CO)}_4]_2}$

$$O{=}CH(CH_2)_6CH_3 \quad \text{(eq 9-9)}$$

9-6. *a)* 2-Pentanone has the CH_3C—R structural feature which will
　　　　　　　　　　　　　　　　　　　‖
　　　　　　　　　　　　　　　　　　　O

give a positive haloform reaction, whereas 3-pentanone does not (sec 9.7b).

b) Any test which distinguishes aldehydes from ketones will work; the Tollens' silver mirror test is one (sec 9.6a).

c) The Tollens' test will be positive for benzaldehyde, but is negative for alcohols.

d) 2-Cyclopentenone has a carbon-carbon double bond; it will therefore decolorize bromine (sec 3.4a) whereas cyclopentanone will not.

e) Ethanol gives a positive haloform reaction, because of the group CH_3CH-R (R=H); methanol does not (sec 9.7b).
|
OH

9-7. a)

$$CH_3-\overset{\overset{\text{O}}{\|}}{C}-CH_3 + H^+ \rightleftharpoons CH_3-\overset{\overset{+\text{O}-H}{\|}}{C}-CH_3$$

$$\Big\Updownarrow H\ddot{O}CH_2CH_2OH$$

$$CH_3-\overset{:\ddot{O}H}{\underset{CH_3}{\overset{|}{C}}}-OCH_2CH_2OH \xrightleftharpoons{-H^+} CH_3-\overset{OH}{\underset{CH_3\ \ H}{\overset{|}{C}}}-\overset{+}{\ddot{O}}CH_2CH_2OH$$

$$H^+ \Big\Updownarrow$$

$$CH_3-\overset{\overset{+}{\ddot{H}\ddot{O}H}}{\underset{CH_3}{\overset{|}{C}}}-OCH_2CH_2OH \xrightleftharpoons{-H_2O} CH_3-\overset{+}{\underset{CH_3}{\overset{|}{C}}}-OCH_2CH_2\ddot{O}H$$

$$\Big\Updownarrow$$

$$CH_3-\overset{O-CH_2}{\underset{CH_3}{\overset{|\ \ \ \ \ \ \ \ }{C}}}\overset{CH_2}{\underset{}{}} \xrightleftharpoons{-H^+} CH_3-\overset{\overset{H}{+}\overset{CH_2}{}}{\underset{CH_3}{C-O}}CH_2$$

b) $CH_3CH-\ddot{O}CH_3 + H^+ \rightleftharpoons CH_3CH-\overset{+}{\ddot{O}}CH_3$
| |
OCH_3 OCH_3

$$\Big\Updownarrow$$

$$CH_3CH-\overset{\overset{H}{|+}}{O}-H \xrightleftharpoons{H\ddot{O}H} CH_3\overset{+}{C}H + CH_3OH$$
$$\underset{OCH_3}{} \qquad\qquad \underset{OCH_3}{}$$

$$-H^+ \Big\Updownarrow \qquad\qquad \Big\Updownarrow H^+$$

$$CH_3CH-OH \rightleftharpoons CH_3CH-OH$$
$$\underset{:\ddot{O}CH_3}{} \qquad\qquad \underset{H\ddot{O}CH_3}{+}$$

$$\Big\Updownarrow$$

$$CH_3CH=O \xrightleftharpoons{-H^+} CH_3\overset{+}{C}H-\ddot{O}-H + CH_3OH$$

c) After protonation and then loss of the first mole of methanol, the acetaldehyde dimethylacetal gives the carbonium ion $CH_3\overset{+}{C}HOCH_3$ (see answer to part b). This ion is resonance stabilized:

$$CH_3\overset{+}{C}H\overset{\frown}{\underset{\cdot\cdot}{\ddot{O}}}CH_3 \longleftrightarrow CH_3CH=\overset{+}{\underset{\cdot\cdot}{\ddot{O}}}CH_3$$

The acid-catalyzed cleavage of methyl sec-butyl ether proceeds as follows:

$$CH_3\underset{\underset{CH_2CH_3}{|}}{CH}-\overset{\cdot\cdot}{\underset{\cdot\cdot}{\ddot{O}}}CH_3 \overset{H^+}{\rightleftharpoons} CH_3\underset{\underset{CH_2CH_3}{|}}{CH}-\overset{\overset{H}{|}}{\underset{\cdot\cdot}{\overset{+}{O}}}CH_3$$

$$\Big\Downarrow$$

$$CH_3\underset{\underset{CH_2CH_3}{|}}{CH^+} \quad + \quad CH_3OH$$

The resulting sec-butyl cation is analogous to the ion from acetaldehyde dimethylacetal except that the oxygen is replaced by a —CH_2— group. With no nearby unshared electron pair, resonance stabilization analogous to that described above is not possible. Therefore acid cleavage of ordinary ethers is more difficult than that of acetals. This fact is frequently useful in studying the chemistry of carbohydrates (Chapter 16).

d) $CH_3CH=O + CH_3CH_2\overset{\cdot\cdot}{\overset{\cdot}{S}}H \rightleftharpoons CH_3\underset{\underset{H}{|}}{CH}-\overset{\overset{O^-}{|}}{\underset{\cdot\cdot}{\overset{+}{S}}}CH_2CH_3$

$$\Big\Updownarrow {\scriptstyle -H^+}$$

$$CH_3\underset{\underset{}{|}}{CH}-SCH_2CH_3 \overset{H^+}{\rightleftharpoons} CH_3\underset{\underset{}{|}}{CH}-SCH_2CH_3$$
$$\overset{+\underset{}{O}H_2}{} \qquad\qquad \overset{OH}{}$$

$$\Big\Updownarrow {\scriptstyle -H_2O}$$

$$CH_3\overset{+}{C}H-SCH_2CH_3 \underset{\text{H}\ddot{\text{S}}CH_2CH_3}{\rightleftharpoons} \begin{array}{c} H-\overset{\cdot\cdot+}{\underset{}{S}}-CH_2CH_3 \\ | \\ CH_3CHSCH_2CH_3 \end{array}$$

$$\Big\Updownarrow {\scriptstyle -H^+}$$

$$CH_3CH(SCH_2CH_3)_2$$

9-8. *a)*

$$CH_3\overset{\displaystyle O}{\overset{\|}{C}}CH_3 + NH_2OH \rightleftharpoons CH_3-\underset{\underset{CH_3}{|}}{\overset{\overset{\textstyle :\ddot{O}:^-}{|}}{C}}-\underset{\underset{H}{|}}{\overset{\overset{H}{|}}{\overset{+}{N}}}-OH$$

$$\Big\updownarrow -H^+$$

$$CH_3-\underset{\underset{CH_3}{|}}{\overset{\overset{\textstyle :\ddot{O}H}{|}}{C}}-NHOH \;\overset{+\,H^+}{\rightleftharpoons}\; CH_3-\underset{\underset{CH_3}{|}}{\overset{\overset{\textstyle O^-}{|}}{C}}-NHOH$$

$$\Big\updownarrow +H^+$$

$$CH_3-\underset{\underset{CH_3}{|}}{\overset{\overset{\textstyle H-\overset{+}{\underset{\ddot{}}{O}}-H}{|}}{C}}-NHOH \;\overset{-H_2O}{\rightleftharpoons}\; CH_3-\underset{\underset{CH_3}{|}}{\overset{+}{C}}-\ddot{N}-OH$$

$$\Big\downarrow -H^+$$

$$CH_3-\underset{\underset{CH_3}{|}}{\overset{}{C}}=NOH$$

$$\text{C}_6\text{H}_5-CH=\ddot{O}: \;+\; \ddot{N}H_2NH-\text{C}_6\text{H}_5 \rightleftharpoons \text{C}_6\text{H}_5-\underset{\underset{H}{|}}{\overset{\overset{\textstyle :\ddot{O}:^-\;H}{|}}{CH}}-\overset{+}{N}-NH-\text{C}_6\text{H}_5$$

$$\Big\updownarrow -H^+$$

$$\text{C}_6\text{H}_5-\overset{\overset{\textstyle :\ddot{O}H\;\;H}{|}}{CH}-\ddot{N}-NH-\text{C}_6\text{H}_5 \;\overset{+H^+}{\rightleftharpoons}\; \text{C}_6\text{H}_5-\overset{\overset{\textstyle O^-\;\;H}{|}}{CH}-\overset{|}{N}-NH-\text{C}_6\text{H}_5$$

$$\Big\updownarrow +H^+$$

$$\text{C}_6\text{H}_5-\overset{\overset{\textstyle H-\overset{+}{\underset{\ddot{}}{O}}-H\;\;H}{|}}{CH}-\ddot{N}-NH-\text{C}_6\text{H}_5 \;\overset{-H_2O}{\rightleftharpoons}\; \text{C}_6\text{H}_5-\overset{+}{CH}-\overset{\overset{H}{|}}{\underset{\ddot{}}{N}}-NH-\text{C}_6\text{H}_5$$

$$\Big\downarrow -H^+$$

$$\text{C}_6\text{H}_5-CH=N-NH-\text{C}_6\text{H}_5$$

b) Acid can catalyze the reaction by protonating the ketone (eq 9-21). However, if the medium is too acidic, the reaction rate will be retarded because of protonation of the hydroxylamine:

$$\ddot{N}H_2\!-\!OH + H^+ \rightleftharpoons H\!-\!\overset{\displaystyle H}{\underset{\displaystyle H}{\overset{|}{\underset{|}{N^+}}}}\!-\!OH$$

In its protonated form, hydroxylamine no longer has an unshared electron pair on the nitrogen; it therefore cannot function as a nucleophile to attack the carbon-oxygen double bond.

9-9. The general equations are 9-33 and 9-34.

a) $CH_3CH\!=\!O + CH_3MgBr \longrightarrow$ $CH_3\overset{\displaystyle O^-\ \overline{MgBr}}{\underset{|}{CH}}\!-\!CH_3$

$\downarrow$ H$^+$, H$_2$O

$CH_3\overset{\displaystyle OH}{\underset{|}{CH}}CH_3$

b)

$\xrightarrow[\text{2. H}_2\text{O, H}^+]{\text{1. CH}_3\text{MgBr}}$

c) $CH_2\!=\!O \xrightarrow[\text{2. H}_2\text{O, H}^+]{\text{1. CH}_3\text{MgBr}} CH_3CH_2OH$

d)

$\xrightarrow[\text{2. H}_2\text{O, H}^+]{\text{1. CH}_3\text{MgBr}}$

9-10. In each case write out the structure of the alcohol:

$$R_1\!-\!\overset{\displaystyle R_2}{\underset{\displaystyle R_3}{\overset{|}{\underset{|}{C}}}}\!-\!O\!-\!H$$

One of the R groups comes from the Grignard reagent; the rest of the molecule comes from the carbonyl compound. For example, if we select R_1 as the alkyl group to be derived from the Grignard reagent, then the carbonyl compound will be $R_2\!-\!\overset{\displaystyle O}{\overset{\|}{C}}\!-\!R_3$.

a) 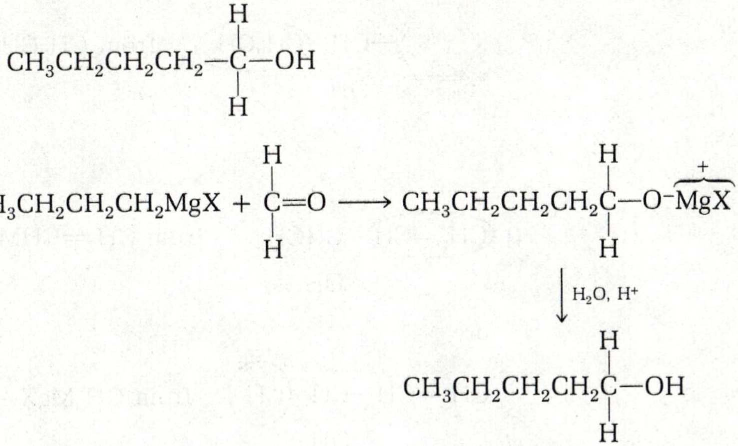 CH$_3$CH$_2$CH$_2$CH$_2$—C—OH (with H above and H below the C)

CH$_3$CH$_2$CH$_2$CH$_2$MgX + C=O (with H above and H below) ⟶ CH$_3$CH$_2$CH$_2$CH$_2$C—O$^-$MgX$^+$ (with H above and H below the C)

↓ H$_2$O, H$^+$

CH$_3$CH$_2$CH$_2$CH$_2$C—OH (with H above and H below the C)

In the remaining cases we will not write out the equations, but will simply show how the initial reactants are derived.

b) 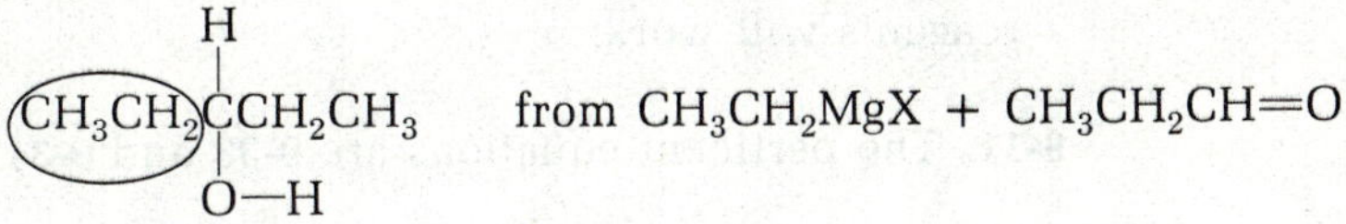 CH$_3$CH$_2$CCH$_2$CH$_3$ (with H above and O—H below the C) from CH$_3$CH$_2$MgX + CH$_3$CH$_2$CH=O

c) CH$_3$—C—CH$_2$CH$_3$ (with CH$_3$ above and OH below the C) from CH$_3$MgX + CH$_3$CCH$_2$CH$_3$ (with O double-bonded below the second C)

CH$_3$—C—CH$_2$CH$_3$ (with CH$_3$ above and OH below the C) from CH$_3$CH$_2$MgX + CH$_3$CCH$_3$ (with O double-bonded below the second C)

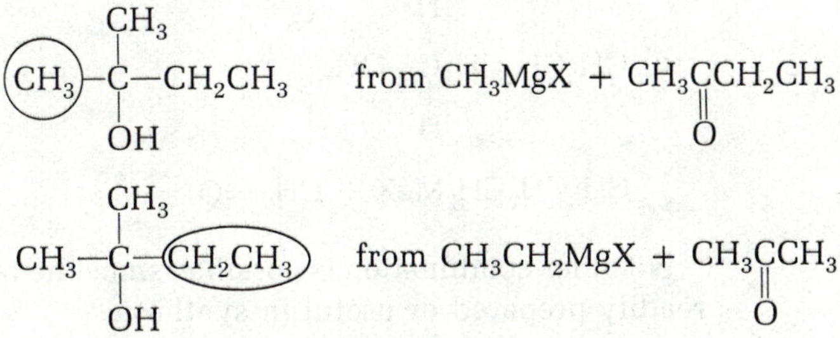

Either combination of reagents will work.

d)

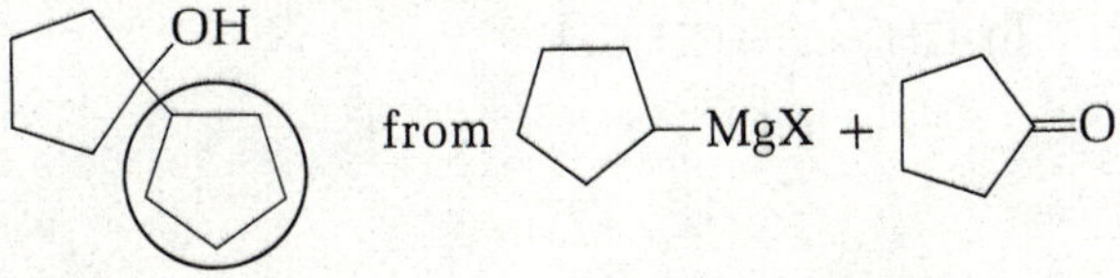

from (cyclopentyl)—MgX + (cyclopentanone) =O

In this case, the "free-standing" R group is selected for the Grignard reagent.

e) 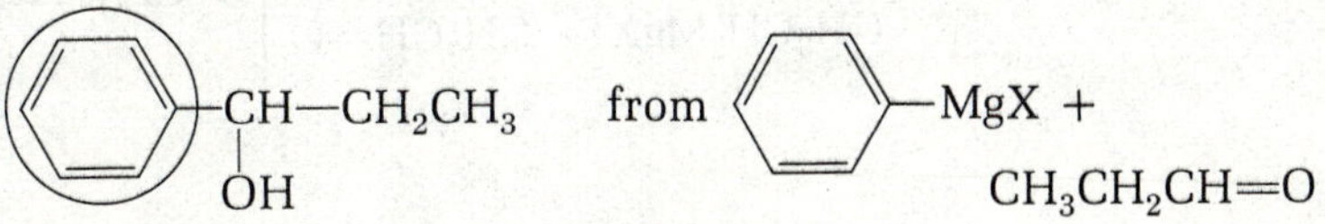 (phenyl)—CH—CH$_2$CH$_3$ (with OH below the CH) from (phenyl)—MgX + CH$_3$CH$_2$CH=O

or

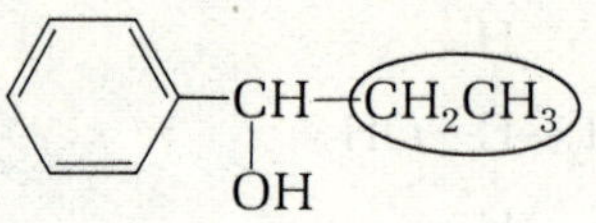

$$\text{C}_6\text{H}_5\text{-CH-CH}_2\text{CH}_3 \quad \text{from } CH_3CH_2MgX +$$
$$\underset{\text{OH}}{|}$$

$$\text{C}_6\text{H}_5\text{-CH=O}$$

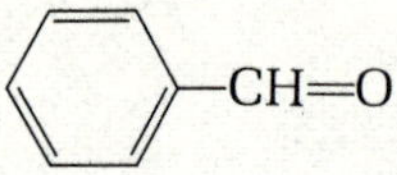

f) $CH_2{=}CH\text{-CHCH}_3 \quad \text{from } CH_2{=}CHMgX + CH_3CH{=}O$
$$\underset{\text{OH}}{|}$$

or

$$CH_2{=}CH\text{-CH-CH}_3 \quad \text{from } CH_3MgX +$$
$$\underset{\text{OH}}{|}$$

$$CH_2{=}CH\text{-CH=O}$$

Vinyl Grignard reagents, though a bit more difficult to prepare than simple alkyl Grignard reagents, are known. Either pair of reagents will work.

9-11. The pertinent equations are 9-33 and 9-34.

a) $$CH_3CH_2CH_2\overset{\displaystyle H}{\underset{\displaystyle H}{\text{C}}}\text{-OH}$$

$$CH_3CH_2CH_2MgX + CH_2{=}O$$

No other combination is possible, since the reagent HMgX is not readily prepared or useful in synthesis.

b) $$CH_3\text{-}\overset{\displaystyle H}{\underset{\displaystyle OH}{\text{C}}}\text{-CH}_2CH_3$$

$$\left.\begin{array}{l} CH_3MgX + CH_3CH_2CH{=}O \\ CH_3CH_2MgX + CH_3CH{=}O \end{array}\right\} \to \overset{\displaystyle O^- \;\overset{+}{MgX}}{CH_3\underset{|}{\text{C}}HCH_2CH_3} \xrightarrow[H^+]{H_2O}$$

$$\overset{\displaystyle OH}{CH_3\underset{|}{\text{C}}HCH_2CH_3}$$

c)

$$CH_3MgX + CH_3CH_2\overset{\overset{\displaystyle O}{\|}}{C}CH_2CH_2CH_3$$

$$CH_3CH_2MgX + CH_3\overset{\overset{\displaystyle O}{\|}}{C}CH_2CH_2CH_3 \longrightarrow CH_3-\overset{\overset{\displaystyle O^- \; \overset{+}{MgX}}{|}}{\underset{\underset{\displaystyle CH_2CH_3}{|}}{C}}-CH_2CH_2CH_3 \xrightarrow[H^+]{H_2O}$$

$$CH_3CH_2CH_2MgX + CH_3\overset{\overset{\displaystyle O}{\|}}{C}CH_2CH_3$$

$$CH_3-\overset{\overset{\displaystyle OH}{|}}{\underset{\underset{\displaystyle CH_2CH_3}{|}}{C}}-CH_2CH_2CH_3$$

9-12. a) $R-\overset{\underset{\displaystyle OH}{|}}{C}H-R$

The two R's must be identical; an example would be $CH_3\underset{\underset{\displaystyle OH}{|}}{C}HCH_3$
from $CH_3MgX + CH_3CHO$

b) $R-\overset{\overset{\displaystyle R}{|}}{\underset{\underset{\displaystyle OH}{|}}{C}}-R$

All three R's must be the same; an example would be $(CH_3)_3COH$

from $CH_3MgBr + CH_3\overset{\overset{\displaystyle O}{\|}}{C}CH_3$

9-13.

$$R-\overset{\overset{\displaystyle O}{\|}}{C}-H + \left[:\overset{\overset{\displaystyle O^-}{|}}{\underset{\underset{\displaystyle O_-}{|}}{\overset{+}{S}}}-OH \right]^- Na^+ \longrightarrow \left[R-\overset{\overset{\displaystyle O^-}{|}}{\underset{\underset{\displaystyle H}{|}}{C}}-\overset{\overset{\displaystyle O^-}{|}}{\underset{\underset{\displaystyle O_-}{|}}{\overset{2+}{S}}}OH \right]^- Na^+$$

$$\Big\updownarrow -H^+$$

$$\left[R-\overset{\overset{\displaystyle OH}{|}}{\underset{\underset{\displaystyle H}{|}}{C}}-\overset{\overset{\displaystyle O^-}{|}}{\underset{\underset{\displaystyle O_-}{|}}{\overset{2+}{S}}}O^- \right] Na^+ \underset{+H^+}{\overset{}{\rightleftharpoons}} \left[R-\overset{\overset{\displaystyle O^-}{|}}{\underset{\underset{\displaystyle H}{|}}{C}}-\overset{\overset{\displaystyle O^-}{|}}{\underset{\underset{\displaystyle O_-}{|}}{\overset{2+}{S}}}O^- \right]^{2-} Na^+$$

The feature of interest is that the sulfur undergoes a change in
formal charge, from 1+ to 2+.

9-14. *a*) An aldol condensation occurs (sec 9.7a).

$$2CH_3CH_2CH{=}O \xrightarrow[\text{heat}]{OH^-}$$

$$\underset{\underset{CH_3}{|}}{CH_3CH_2\overset{\overset{OH}{|}}{C}HCHCH{=}O} \qquad \text{(eq 9-49; } R = CH_3)$$

b) Halogenation occurs, α to the carbonyl group (sec 9.7b).

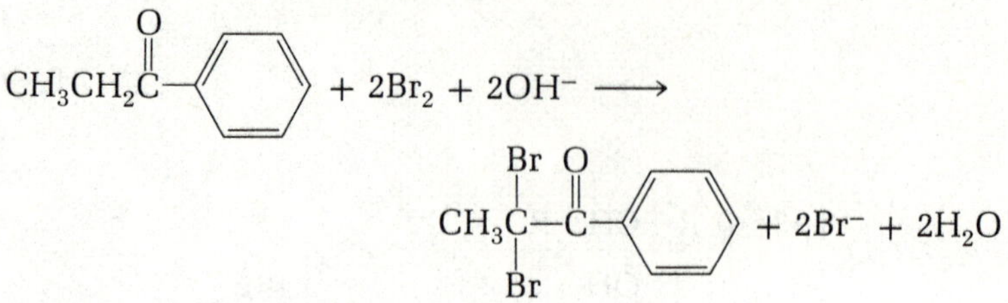

$$CH_3CH_2\overset{\overset{O}{\|}}{C}{-}\bigcirc + 2Br_2 + 2OH^- \longrightarrow$$

$$\underset{\underset{Br}{|}}{CH_3\overset{\overset{Br}{|}}{C}{-}}\overset{\overset{O}{\|}}{C}{-}\bigcirc + 2Br^- + 2H_2O$$

c) The second halogen enters preferentially on the same "side" of the carbonyl group as the first:

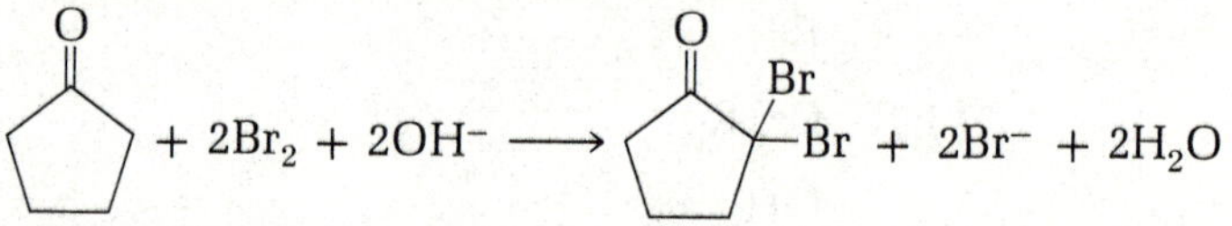

$$\bigcirc\!\!=\!\!O + 2Br_2 + 2OH^- \longrightarrow \bigcirc\!\!\overset{O}{<}\!\!\overset{Br}{\underset{Br}{}} + 2Br^- + 2H_2O$$

The other dibromo product, $Br{-}\bigcirc{-}Br$, is a minor by-product.

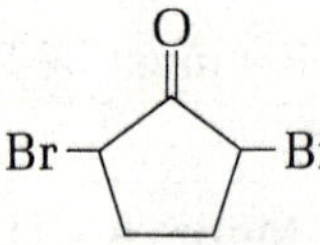

d)

$$\bigcirc\!\!=\!\!O \xrightarrow{LiAlH_4} \bigcirc\!\!\overset{H}{<}\!\!\overset{}{OH}$$

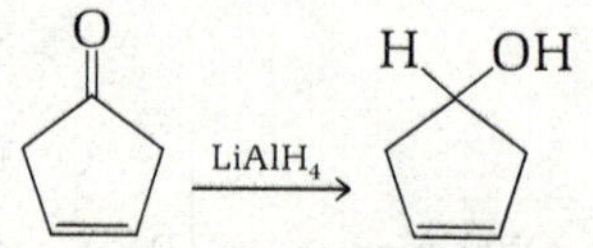

The carbonyl group is reduced without reduction of the carbon-carbon double bond (eq 9-43).

e)

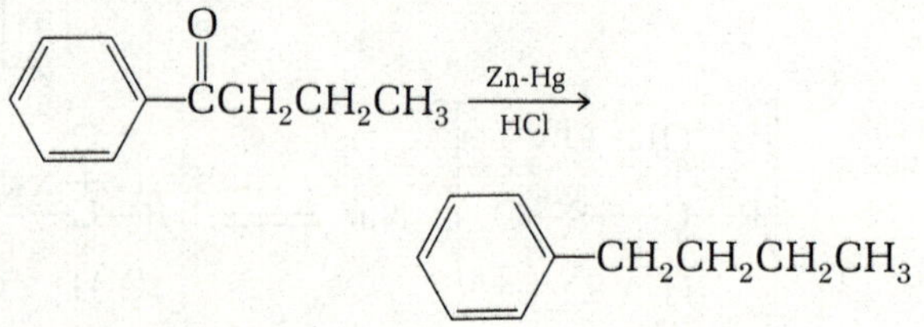

$$\bigcirc{-}\overset{\overset{O}{\|}}{C}CH_2CH_2CH_3 \xrightarrow[\text{HCl}]{\text{Zn-Hg}}$$

$$\bigcirc{-}CH_2CH_2CH_2CH_3 \qquad \text{(eq 9-44)}$$

9-15. Review section 9.7.

The four hydrogens α to the carbonyl group are exchanged; the structure of the product is shown below:

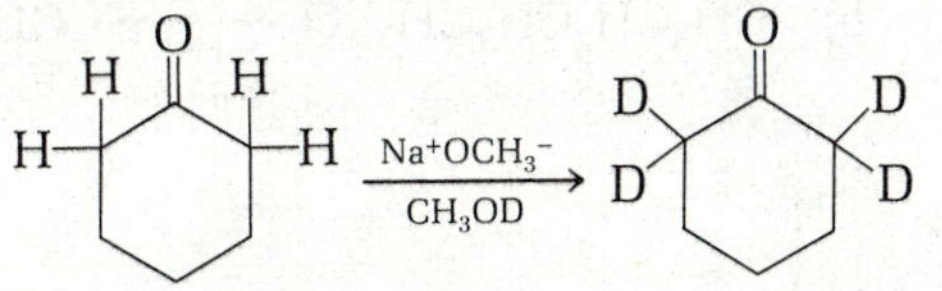

The mechanism involves the reversible formation of the enolate anion (eq 9-45).

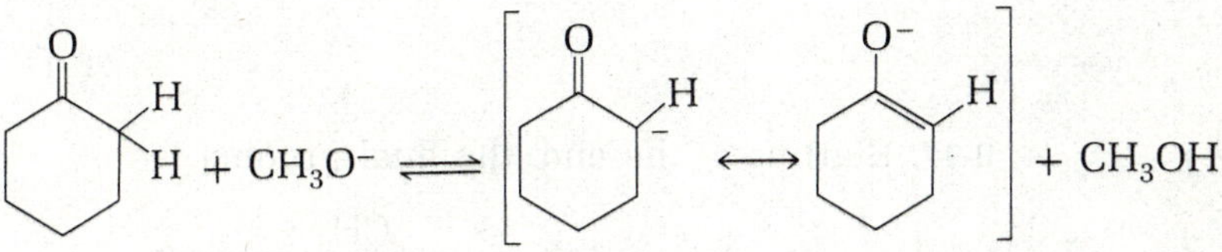

Since CH_3OD is present in large excess, reprotonation of the anion introduces deuterium:

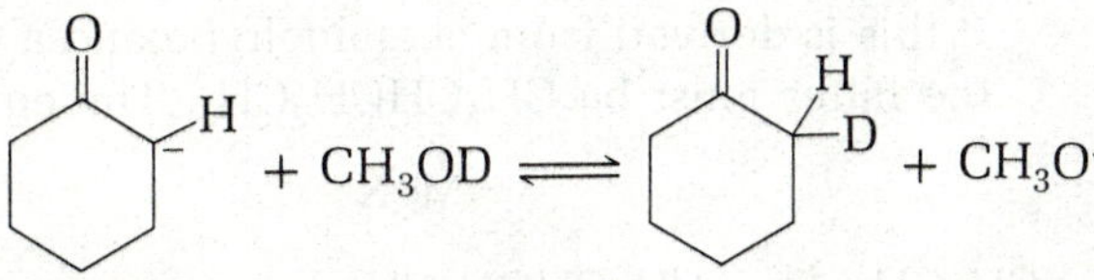

Repetition of this process leads to replacement of the remaining H's on the α-carbons. Since enolization involves only the α-protons, only these are exchanged. The four hydrogens at carbons 2 and 6 are exchanged, whereas the hydrogens attached to carbons 3, 4 and 5 are unaffected.

This is a useful way of introducing deuterium in specific positions in an organic molecule.

9-16. *a*) $RCH_2\overset{O}{\overset{\|}{C}}H + OH^- \rightleftharpoons R\overset{-}{C}H\overset{O}{\overset{\|}{C}}H + H_2O$

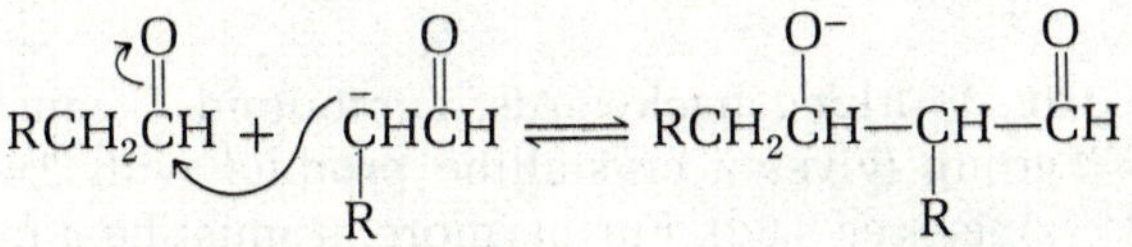

$$RCH_2\overset{O^-}{\overset{|}{C}}H\underset{R}{\underset{|}{C}}HCHO + H_2O \rightleftharpoons RCH_2\overset{OH}{\overset{|}{C}}H\underset{R}{\underset{|}{C}}HCHO + OH^-$$

b) 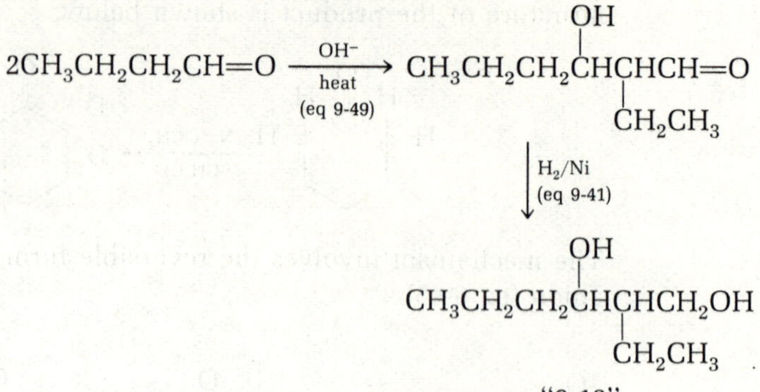 $2CH_3CH_2CH_2CH{=}O \xrightarrow[\text{heat}]{OH^-}$
$\underset{\text{(eq 9-49)}}{}$
$CH_3CH_2CH_2\overset{\displaystyle OH}{\underset{\displaystyle \underset{CH_2CH_3}{|}}{\overset{|}{C}}HCHCH{=}O}$

$\Big\downarrow \begin{array}{l}H_2/Ni\\ \text{(eq 9-41)}\end{array}$

$CH_3CH_2CH_2\overset{\displaystyle OH}{\underset{\displaystyle \underset{CH_2CH_3}{|}}{\overset{|}{C}}HCHCH_2OH}$.

"6-12"

9-17. Starting at the end, the final product is

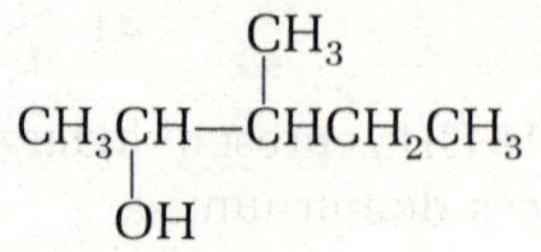

$$CH_3\underset{\displaystyle \underset{OH}{|}}{CH}{-}\overset{\displaystyle \overset{CH_3}{|}}{C}HCH_2CH_3$$

If this is derived from acetaldehyde and a Grignard reagent, then the latter must be $CH_3\underset{\displaystyle \underset{MgX}{|}}{C}HCH_2CH_3$. The equation would be:

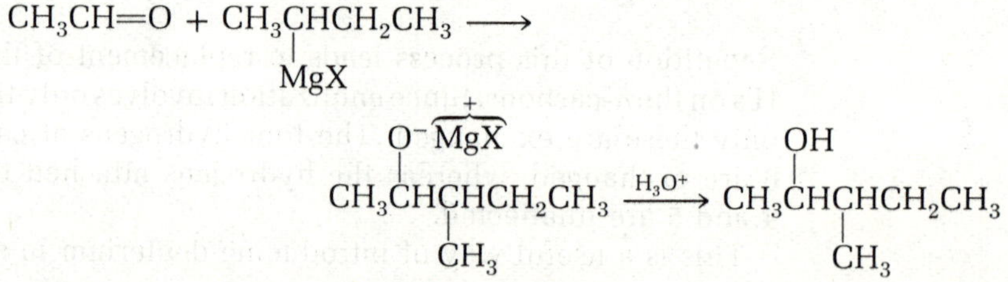

$$CH_3CH{=}O + CH_3\underset{\displaystyle \underset{MgX}{|}}{C}HCH_2CH_3 \longrightarrow$$

$CH_3\overset{\displaystyle \overset{O^-\overset{+}{MgX}}{|}}{C}H\underset{\displaystyle \underset{CH_3}{|}}{C}HCH_2CH_3 \xrightarrow{H_3O^+} CH_3\overset{\displaystyle \overset{OH}{|}}{C}H\underset{\displaystyle \underset{CH_3}{|}}{C}HCH_2CH_3$

Therefore the original alkyl bromide must have been 2-bromobutane, and the remaining equations are:

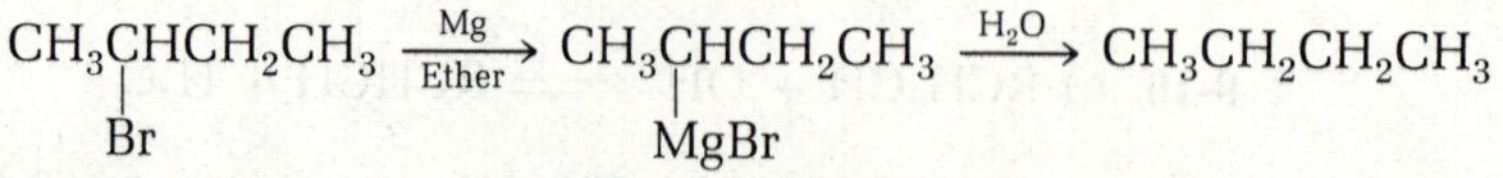

$CH_3\underset{\displaystyle \underset{Br}{|}}{C}HCH_2CH_3 \xrightarrow[\text{Ether}]{Mg} CH_3\underset{\displaystyle \underset{MgBr}{|}}{C}HCH_2CH_3 \xrightarrow{H_2O} CH_3CH_2CH_2CH_3$

9-18. Working backwards, compound C must contain a carbonyl group (gives a crystalline product with 2,4-dinitrophenylhydrazine—sec 9.6d). Furthermore, it must be a methyl ketone, since it gives a positive iodoform test. Therefore C must have the partial structure

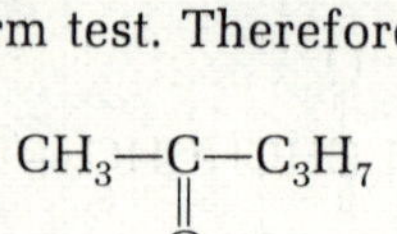

$$CH_3{-}\overset{\displaystyle }{\underset{\displaystyle \underset{O}{\|}}{C}}{-}C_3H_7$$

Two possibilities are

$$CH_3-\underset{\underset{O}{\|}}{C}-CH_2CH_2CH_3 \quad \text{and} \quad CH_3-\underset{\underset{O}{\|}}{C}-CH(CH_3)_2$$

Compound B must therefore be the corresponding secondary alcohol, since it is oxidized to C with chromic acid (sec 9.4a). Possibilities for B are

$$CH_3\underset{\underset{OH}{|}}{C}HCH_2CH_2CH_3 \quad \text{and} \quad CH_3\underset{\underset{OH}{|}}{C}HCH(CH_3)_2$$

Since B is formed from CH_3MgBr + A, A must be either

$$O{=}CHCH_2CH_2CH_3 \quad \text{or} \quad O{=}CHCH(CH_3)_2$$

Insufficient information is given to make a more complete assignment of the structures. Selecting A = *n*-butyraldehyde, the equations for the reactions are:

$$CH_3CH_2CH_2CH{=}O + CH_3MgBr \longrightarrow CH_3CH_2CH_2\overset{\overset{\displaystyle O^-\overset{+}{MgBr}}{|}}{C}HCH_3$$

$$\Big\downarrow {H_3O^+}$$

$$CH_3CH_2CH_2\overset{\overset{\displaystyle O}{\|}}{C}CH_3 \xleftarrow[H^+]{CrO_3} CH_3CH_2CH_2\overset{\overset{\displaystyle OH}{|}}{C}HCH_3$$

$$CH_3CH_2CH_2\overset{\overset{\displaystyle O}{\|}}{C}CH_3 + NH_2NH{-}\!\!\!\bigcirc\!\!\!{-}NO_2 \longrightarrow$$

(with NO_2 substituent on ring)

$$\overset{\overset{\displaystyle N-NH-\!\!\bigcirc\!\!-NO_2}{\|}}{CH_3CH_2CH_2CCH_3} + H_2O$$

(with NO_2 substituent on ring)

crystalline 2,4-dinitrophenylhydrazone

$$CH_3CH_2CH_2\overset{\overset{\displaystyle O}{\|}}{C}CH_3 + 3I_2 + 4OH^- \xrightarrow[\text{(eq 9-54)}]{}$$

$$CHI_3 + CH_3CH_2CH_2CO_2^- + 3I^- + 3H_2O$$

iodoform

9-19. In each case, write the structure of the desired product and work backwards.

a) $CH_3CH_2CH_2CH{=}\underset{\underset{CH_2CH_3}{|}}{C}{-}CH{=}O \xleftarrow[\substack{H^+\\ \text{(eq 9-50;}\\ \text{also, sec 3.5a)}}]{H_2SO_4}$

$CH_3CH_2CH_2\underset{\underset{CH_2CH_3}{|}}{\overset{\overset{OH}{|}}{CH}}{-}CH{-}CH{=}O$

$\Big\uparrow \substack{OH^- \\ \text{(eq 9-49)}}$

$2CH_3CH_2CH_2CH{=}O$

b) $CH_3CH_2\overset{\overset{OH}{|}}{CH}{-}\underset{\underset{CH_3}{|}}{CH}{-}CH_2OH \xleftarrow[\text{(eq 9-41)}]{H_2/Ni}$

$CH_3CH_2\overset{\overset{OH}{|}}{CH}{-}\underset{\underset{CH_3}{|}}{CH}{-}CH{=}O$

$\Big\uparrow \substack{OH^- \\ \text{(eq 9-49)}}$

$2CH_3CH_2CH{=}O$

c) $\langle\!\!\langle\text{benzene}\rangle\!\!\rangle{-}CH_2CH_2CH_2OH \xleftarrow[\text{Ni}]{2H_2} \langle\!\!\langle\text{benzene}\rangle\!\!\rangle{-}CH{=}CH{-}CH{=}O$

cinnamaldehyde

$\Big\uparrow \substack{OH^- \\ \text{(eq 9-51)}}$

$\langle\!\!\langle\text{benzene}\rangle\!\!\rangle{-}CH{=}O + CH_3CH{=}O$

d) $CH_3{-}\underset{\underset{CH_3}{|}}{\overset{\overset{OH}{|}}{C}}{-}CH_2\overset{\overset{O}{\|}}{C}CH_3 \xleftarrow[\text{(eq 9-49)}]{OH^-} 2CH_3\overset{\overset{O}{\|}}{C}CH_3$

In this case the starting material is a ketone rather than an aldehyde. The equilibrium favors the ketone rather than its dimer (a ketol), but if special techniques are used, good yields of the ketol can be obtained.

e) 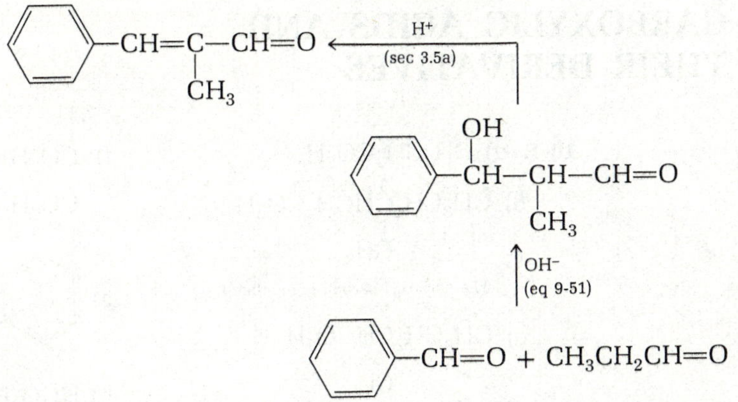 (benzene ring)$-CH=\underset{\underset{CH_3}{|}}{C}-CH=O \xleftarrow[\text{(sec 3.5a)}]{H^+}$

(benzene ring)$-CH-\underset{\underset{CH_3}{|}}{CH}-CH=O$ (with OH above first CH)

$\uparrow$ OH^-
(eq 9-51)

(benzene ring)$-CH=O + CH_3CH_2CH=O$

9-20. The fact that deuterium exchange occurs at the methyl group suggests that the enolate anion must be formed by removal of a methyl proton (compare with problem 9-15). The reason for this is that the ion can be resonance stabilized:

$$CH_3CH=CHCH=O + {}^-OCH_3 \rightleftharpoons$$
$$^-CH_2-CH=CH-CH=O + CH_3OH$$

$$\bar{C}H_2-CH=CH-CH=O \longleftrightarrow CH_2=CH-CH=CH-O^-$$

If reprotonation occurs, using CH_3OD, deuterium will be introduced on the methyl carbon:

$$^-CH_2-CH=CH-CH=O + CH_3OD \rightleftharpoons$$
$$D-CH_2-CH=CH-CH=O + CH_3O^-$$

In the presence of a large excess of CH_3OD, all the methyl protons will exchange.

Similarly, the enolate anion can react with benzaldehyde according to the following mechanism:

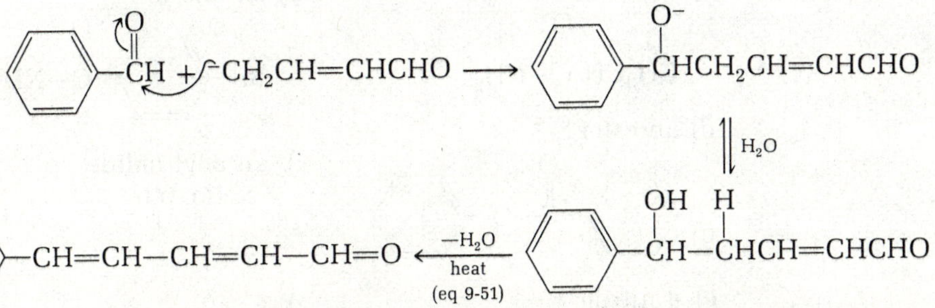

(benzene ring)$-\overset{\overset{O}{\|}}{CH} + {}^-CH_2CH=CHCHO \longrightarrow$ (benzene ring)$-\overset{\overset{O^-}{|}}{CH}CH_2CH=CHCHO$

$\Updownarrow H_2O$

(benzene ring)$-CH=CH-CH=CH-CH=O \xleftarrow[\text{heat}]{-H_2O}$ (benzene ring)$-\overset{\overset{OH}{|}}{CH}-\overset{\overset{H}{|}}{CH}CH=CHCHO$
(eq 9-51)

Dehydration of the aldol occurs readily, since the product has a fully conjugated double bond system.

CHAPTER TEN

CARBOXYLIC ACIDS AND THEIR DERIVATIVES

10-1. a) $CH_3CH_2CO_2H$

b) $CH_3CH_2\overset{3}{C}H\overset{2}{C}H\overset{1}{C}H_2CO_2H$
    |
    CH_3

The carboxyl carbon is number 1.

c) $CH_3CH_2CHCO_2H$
       |
       Cl

d) CH_3--CO_2H

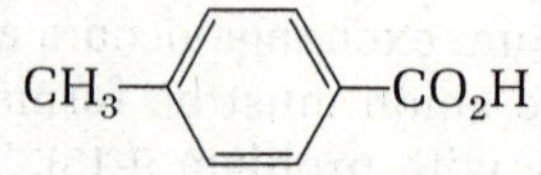

e) 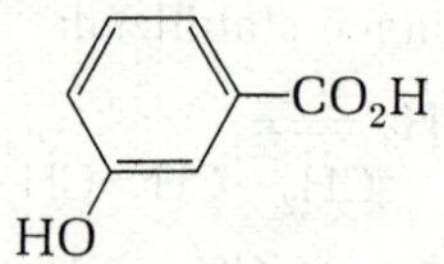 $-CO_2H$; HO

f) CO_2H
   CO_2H

g) 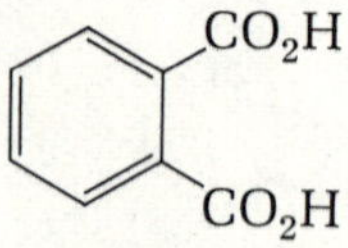 CO_2H ; CO_2H

h) HCO_2H

i) 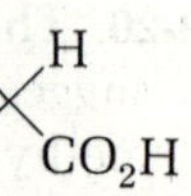 $\overset{H}{CO_2H}$

j) $\overset{\gamma}{C}H_3\overset{\beta}{C}H\overset{\alpha}{C}H_2CO_2H$
        |
        Br

Greek letters begin with the carbon atom adjacent to the carboxyl carbon.

10-2. a) an ester

$CH_3CH_2\overset{O}{\overset{\|}{C}}-OCH_3$

b) an acyl halide

$CH_3CH_2CH_2\overset{O}{\overset{\|}{C}}-Br$

c) an amide

$CH_3CH_2\overset{O}{\overset{\|}{C}}-NH_2$

d) an ester

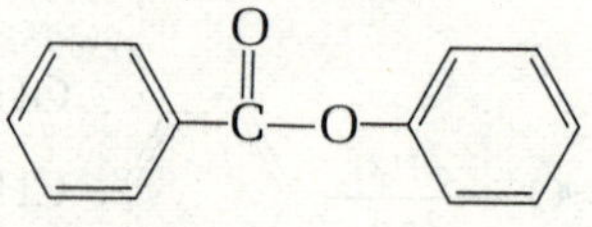

$\overset{O}{\overset{\|}{C}}-O-$

e) a nitrile

$-C\equiv N$

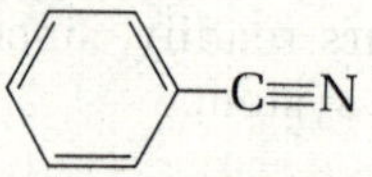

f) an acid anhydride

$CH_3\overset{O}{\overset{\|}{C}}-O-\overset{O}{\overset{\|}{C}}CH_3$

g) a salt

$H\overset{O}{\overset{\|}{C}}-O^-NH_4^+$

h) an amide

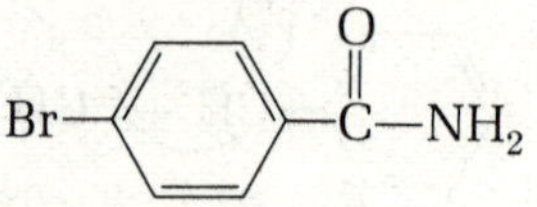

$Br--\overset{O}{\overset{\|}{C}}-NH_2$

i) an acyl halide

$\overset{O}{\overset{\|}{Cl-C}}-\overset{O}{\overset{\|}{C}}-Cl$

j) a salt

$(CH_3\overset{O}{\overset{\|}{C}}-O^-)_2Ca^{2+}$

10-3. *a)* 4-methylpentanoic acid

    b)   3-bromo-2-methylbutanoic acid

    c)   *ortho*-nitrobenzoic acid

    d)   2-phenylpropanoic acid
       (or α-phenylpropionic acid)

    e)   acrylic acid
       (or propenoic acid)

    f)   cyclohexanecarboxylic acid

    g)   2,2-difluoropropanoic acid
       (or α,α-difluoropropionic acid)

    h)   2-naphthoic acid
       (or β-naphthoic acid)

    i)   3-methylpentanedioic acid
       (or 3- or β-methylglutaric acid)

    j)   tetrachloroterephthalic acid

10-4. Notice that all the names of the acids in question 10-3 end with the word acid; none of the names of acid derivatives, given in this question, end with that word.

    a) calcium butyrate
       (or calcium butanoate)

    b) phenyl isobutyrate
       (or phenyl 2-methylpropanoate)

    c) propionyl chloride
       (or propanoyl chloride)

    d) α-chloroacetamide

(The α- is used to distinguish this compound from $CH_3\overset{\text{O}}{\overset{\|}{C}}NHCl$, which is N-chloroacetamide.)

    e) sodium *p*-toluate

    f) benzoic anhydride

    g) methyl trifluoroacetate

    h) ethyl formate

    i) tetrachlorophthalic anhydride

    j) methyl oxalate

10-5. *a)* $\text{C}_6\text{H}_5-CO_2H + Na^+OH^- \longrightarrow \text{C}_6\text{H}_5-CO_2^-Na^+ + H_2O$

                    sodium benzoate

This is a neutralization reaction, equation 10-2.

b) $2 \langle \bigcirc \rangle -CO_2H + 2 K \longrightarrow 2 \langle \bigcirc \rangle -CO_2^-K^+ + H_2$

potassium benzoate

This is analogous to equation 6-4. Normally, salts of acids are not made this way, for the reaction is quite exothermic. But all acids do react with the alkali metals to liberate hydrogen.

c) $2 \langle \bigcirc \rangle -CO_2H + Ca^{2+}(OH)_2^- \xrightarrow[\text{eq 10-2}]{}$

$$\left(\langle \bigcirc \rangle -CO_2^- \right)_2 Ca^{2+} + 2 H_2O$$

calcium benzoate

d) $\langle \bigcirc \rangle -CO_2H + NH_3 \longrightarrow \langle \bigcirc \rangle -CO_2^-NH_4^+ \xrightarrow{\text{heat}}$

$$\langle \bigcirc \rangle -\overset{\overset{\textstyle O}{\|}}{C}-NH_2 + H_2O \qquad \text{(eq 10-45, 10-46)}$$

e) $\langle \bigcirc \rangle -CO_2H + SOCl_2 \xrightarrow[\text{(eq 10-31)}]{}$

$$\langle \bigcirc \rangle -\overset{\overset{\textstyle O}{\|}}{C}-Cl + HCl + SO_2$$

f) $\langle \bigcirc \rangle -CO_2H + HOCH_2CH_3 \xrightarrow[\text{(eq 10-18)}]{H^+}$

$$\langle \bigcirc \rangle -\overset{\overset{\textstyle O}{\|}}{C}-OCH_2CH_3 + H_2O$$

g) $\langle \bigcirc \rangle -CO_2H + HCl \longrightarrow$ no reaction

h) $\langle \bigcirc \rangle -CO_2H + H_2O \longrightarrow$ no reaction, but benzoic acid is quite soluble in hot water, relatively insoluble in cold water. Pure needles can be obtained by recrystallization from water.

10-6. The information necessary to correct answers is given in section 10.2.

  a) CH_2ClCO_2H; both substituents, Cl and Br, are approximately the same distance from the carboxyl group, but Cl is more electronegative than Br.

  b) o-bromobenzoic acid; the bromine is closer to the carboxyl group, and is an electron-withdrawing substituent. Compare the pK_a's of the corresponding chloro acids, given in Table 10-3.

  c) CF_3CO_2H; fluorine is more electronegative than chlorine.

  d) benzoic acid; methoxyl is an electron-releasing substituent, when in the *para* position, and may destabilize the anion due to structures such as

which bring two negative charges near one another.

  e) $CH_3CHClCO_2H$; the chlorine, which is electron-withdrawing, is closer to the carboxyl group.

10-7. *a)* $CH_3CH_2CH_2CH_2OH \xrightarrow[\text{(eq 10-7)}]{Na_2Cr_2O_7} CH_3CH_2CH_2CO_2H$

  b) $CH_3CH_2CH_2OH \xrightarrow[\text{(sec 6.7a)}]{HBr} CH_3CH_2CH_2Br$

  $CH_3CH_2CH_2Br \xrightarrow[\text{(eq 10-12)}]{NaCN} CH_3CH_2CH_2CN \xrightarrow[\substack{H^+ \\ \text{(eq 10-13)}}]{H_2O}$

  $CH_3CH_2CH_2CO_2H$

  $CH_3CH_2CH_2Br \xrightarrow{Mg} CH_3CH_2CH_2MgBr \xrightarrow[2)\ H_2O]{1)\ CO_2}$

  $CH_3CH_2CH_2CO_2H$ (eq 10-15)

  c)

(eq 5-20 or 10-3)

  d)

succinic acid

e)

$$\text{cyclopentane} \xrightarrow[\text{light}]{Cl_2} \text{chlorocyclopentane} \xrightarrow[\begin{array}{l}\text{1) Mg}\\\text{2) }CO_2\\\text{3) }H_2O,\ H^+\end{array}]{} \quad \text{(eq 10-15)}$$

f)

$$CH_2-CH_2 \underset{O}{\ } \xrightarrow[\text{(eq 7-21)}]{\underset{H^+}{CH_3OH}} CH_3OCH_2CH_2OH \xrightarrow[\text{(eq 10-7)}]{Na_2Cr_2O_7}$$

$$CH_3OCH_2CO_2H$$

g)

$$\xrightarrow[\text{heat}]{KMnO_4} \begin{array}{l} HOOC\\ HOOC \end{array} \quad \text{(eq 10-6)}$$

h) $CH_3CH{=}CHCH_3 \xrightarrow[\text{(eq 3-11)}]{HBr} CH_3\underset{Br}{CH}CH_2CH_3 \xrightarrow[\begin{array}{l}\text{1. Mg}\\\text{2. }CO_2\\\text{3. }H^+\end{array}]{\text{(eq 10-15)}}$

$$CH_3\underset{CO_2H}{CH}CH_2CH_3$$

10-8. a) $CH_3CH_2CO_2H \xrightarrow{PBr_3} CH_3CH_2\overset{O}{\overset{\|}{C}}{-}Br \quad \text{(eq 10-29)}$

b) $CH_3CH_2CH_2CH_2CO_2H + CH_3CH_2OH \xrightarrow[\text{heat}]{H^+}$
   (eq 10-18)

$$CH_3CH_2CH_2CH_2\overset{O}{\overset{\|}{C}}{-}OCH_2CH_3 + H_2O$$

c) $CH_3CH_2CH_2CO_2H + NH_4OH \xrightarrow[\text{(eq 10-46)}]{\text{heat}}$

$$CH_3CH_2CH_2\overset{O}{\overset{\|}{C}}NH_2 + 2H_2O$$

d)

$$\xrightarrow{\text{heat}} \quad + H_2O \quad \text{(eq 10-34)}$$

e) $HO_2C{-}CO_2H + Ca(OH)_2 \xrightarrow{\text{(eq 10-2)}}$

$$(^-O_2C{-}CO_2{}^-)Ca^{2+} + 2H_2O$$

f) $\bigcirc$—CH$_2$CO$_2$H + NH$_3$ $\xrightarrow[\text{(eq 10-46)}]{\text{heat}}$

$$\bigcirc-CH_2\overset{\overset{\displaystyle O}{\|}}{C}-NH_2 + H_2O$$

g) CH$_3$CH$_2$CH$_2$CH$_2$CO$_2$H $\xrightarrow[\text{(eq 10-30)}]{\text{PCl}_5}$ CH$_3$CH$_2$CH$_2$CH$_2\overset{\overset{\displaystyle O}{\|}}{C}-$Cl

CH$_3$CH$_2$CH$_2$CH$_2$CO$_2$H $\xrightarrow[\text{(eq 10-2)}]{\text{NaOH}}$ CH$_3$CH$_2$CH$_2$CH$_2$CO$_2^-$Na$^+$

CH$_3$CH$_2$CH$_2$CH$_2\overset{\overset{\displaystyle O}{\|}}{C}-$Cl + CH$_3CH_2CH_2CH_2CO_2^-Na^+$ $\xrightarrow[\text{(eq 10-33)}]{\text{heat}}$

CH$_3$CH$_2$CH$_2$CH$_2\overset{\overset{\displaystyle O}{\|}}{C}-O-\overset{\overset{\displaystyle O}{\|}}{C}CH_2CH_2CH_2CH_3$ + Na$^+$Cl$^-$

h) $\bigcirc$—CO$_2$H + (CH$_3$)$_2$CHOH $\xrightarrow[\text{(eq 10-18)}]{\text{H}^+}$

$$\bigcirc-\overset{\overset{\displaystyle O}{\|}}{C}-O-CH(CH_3)_2 + H_2O$$

i) $\underset{\text{NO}_2}{\overset{\text{CO}_2\text{H}}{\bigcirc}}$ + PCl$_5$ $\xrightarrow[\text{(eq 10-30)}]{}$ $\underset{\text{NO}_2}{\overset{\text{COCl}}{\bigcirc}}$ + HCl + POCl$_3$

j) $H\overset{\overset{\displaystyle O}{\|}}{C}-$OH + NH$_3$ $\xrightarrow[\text{(eq 10-46)}]{\text{heat}}$ $H\overset{\overset{\displaystyle O}{\|}}{C}-NH_2$ + H$_2$O

10-9. a) $R-\overset{\overset{\displaystyle O}{\|}}{C}-OR' + :NH_3 \rightleftharpoons R-\underset{\overset{\displaystyle +}{NH_3}}{\overset{\overset{\displaystyle O^-}{|}}{C}}-OR' \underset{\sim H^+}{\rightleftharpoons} R-\underset{\displaystyle NH_2}{\overset{\overset{\displaystyle OH}{|}}{C}}-OR'$

$$H^+ \Updownarrow$$

$R-\underset{\displaystyle NH_2}{\overset{\overset{\displaystyle O}{\|}}{C}} \underset{-H^+}{\rightleftharpoons} R-\underset{\displaystyle NH_2}{\overset{\displaystyle +}{C}} \underset{-R'OH}{\rightleftharpoons} R-\underset{\displaystyle NH_2}{\overset{\overset{\displaystyle OH\ H}{|\ \ |}}{C}}-\overset{+}{O}R'$

The details of the various protonations and deprotonations are difficult to establish precisely, but the scheme written above is one of several plausible alternatives.

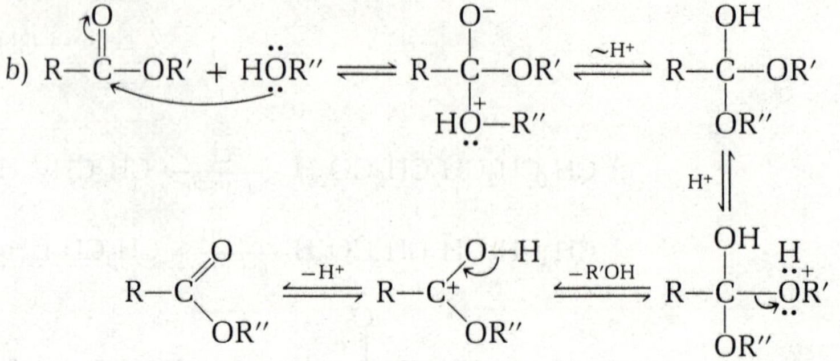

b) $R-\overset{\displaystyle O}{\overset{\|}{C}}-OR' + H\overset{..}{\underset{..}{O}}R'' \rightleftharpoons R-\overset{\displaystyle O^-}{\underset{\underset{..}{H\overset{+}{O}}-R''}{C}}-OR' \underset{\sim H^+}{\rightleftharpoons} R-\overset{\displaystyle OH}{\underset{OR''}{C}}-OR'$

$H^+ \big\Updownarrow$

$R-\overset{O}{\underset{OR''}{C}} \underset{-H^+}{\rightleftharpoons} R-\overset{O-H}{\underset{OR''}{C^+}} \underset{-R'OH}{\longleftarrow} R-\overset{OH}{\underset{OR''}{C}}\overset{H}{\underset{..}{\overset{..+}{O}R'}}$

10-10. a) The first step involves nucleophilic attack on the carbonyl group of the acyl halide.

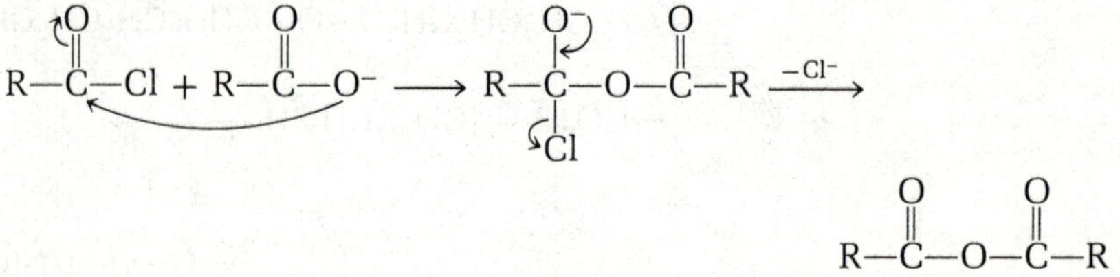

$R-\overset{\displaystyle O}{\overset{\|}{C}}-Cl + R-\overset{\displaystyle O}{\overset{\|}{C}}-O^- \longrightarrow R-\overset{\displaystyle O^-}{\underset{\underset{Cl}{|}}{C}}-O-\overset{\displaystyle O}{\overset{\|}{C}}-R \overset{-Cl^-}{\longrightarrow}$

$R-\overset{\displaystyle O}{\overset{\|}{C}}-O-\overset{\displaystyle O}{\overset{\|}{C}}-R$

b) The first step involves an intramolecular nucleophilic attack. Regardless of which oxygen functions as the nucleophile, the net result is the same:

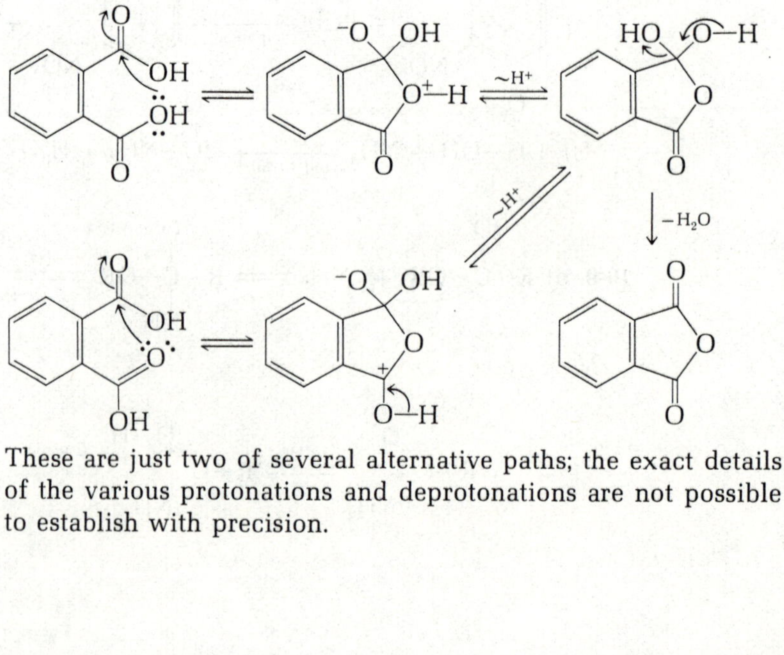

These are just two of several alternative paths; the exact details of the various protonations and deprotonations are not possible to establish with precision.

c) $CH_2=C=O + CH_3\overset{\overset{O}{\|}}{C}-\overset{..}{\underset{..}{O}}H \longrightarrow CH_2=\overset{\overset{O^-}{|}}{C}-\overset{..+}{\underset{H}{O}}-\overset{\overset{O}{\|}}{C}CH_3$

$$\downarrow$$

$CH_3\overset{\overset{O}{\|}}{C}-O-\overset{\overset{O}{\|}}{C}CH_3 \underset{\sim H^+}{\rightleftharpoons} {}^-CH_2-\overset{\overset{O}{\|}}{C}-\overset{..+}{\underset{H}{O}}-\overset{\overset{O}{\|}}{C}CH_3$

Again, the detailed steps of proton addition and removal are not known, but the main point is that reaction is initiated by nucleophilic attack on the ketene carbonyl group.

10-11.

Eq 10-36

$CH_3\overset{\overset{O}{\|}}{C}-Cl + H-\overset{..}{\underset{..}{O}}-H \longrightarrow CH_3\overset{\overset{O}{\|}}{C}-\overset{..+}{\underset{\underset{Cl\ H}{|}}{O}}-H \xrightarrow{-Cl^-} CH_3-\overset{\overset{O}{\|}}{C}-\overset{..+}{\underset{H}{O}}-H$

$\Bigg\updownarrow \sim H^+ \qquad\qquad -H^+ \Bigg\updownarrow +H^+$

$CH_3\overset{\overset{O-H}{|}}{\underset{\underset{Cl}{|}}{C}}-OH \xrightarrow{-HCl} CH_3-\overset{\overset{O}{\|}}{C}-OH$

The details of proton addition and removal are not known. This comment applies to all of the equations in this problem; only one of several plausible paths which follow the initial nucleophilic attack will be shown.

Eq 10-37

$CH_3\overset{\overset{O}{\|}}{C}-O-\overset{\overset{O}{\|}}{C}CH_3 + H\overset{..}{\underset{..}{O}}H \longrightarrow CH_3\overset{\overset{O^-}{|}}{\underset{\underset{\overset{\|}{O}}{OCCH_3}}{C}}-\overset{+..}{O}\overset{H}{\underset{H}{<}} \underset{\sim H^+}{\rightleftharpoons} CH_3\overset{\overset{O-H}{|}}{\underset{\underset{\overset{\|}{O}}{OCCH_3}}{C}}-OH$

$$\downarrow$$

$$2CH_3\overset{\overset{O}{\|}}{C}OH$$

(Attack may occur at either carbonyl group of the anhydride.)

Eq 10-38

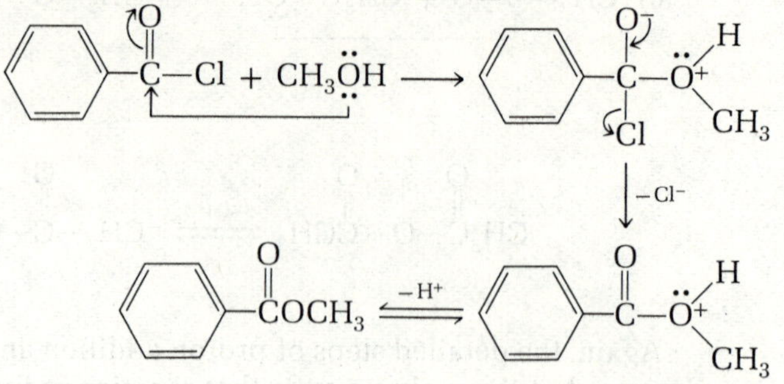

Eq 10-39

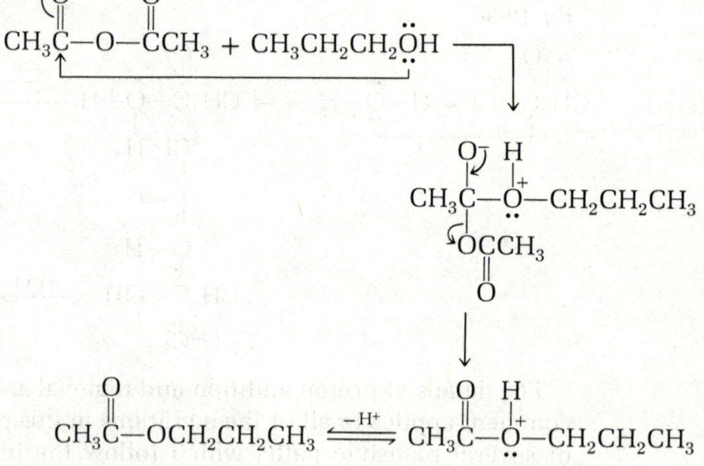

Eq 10-40

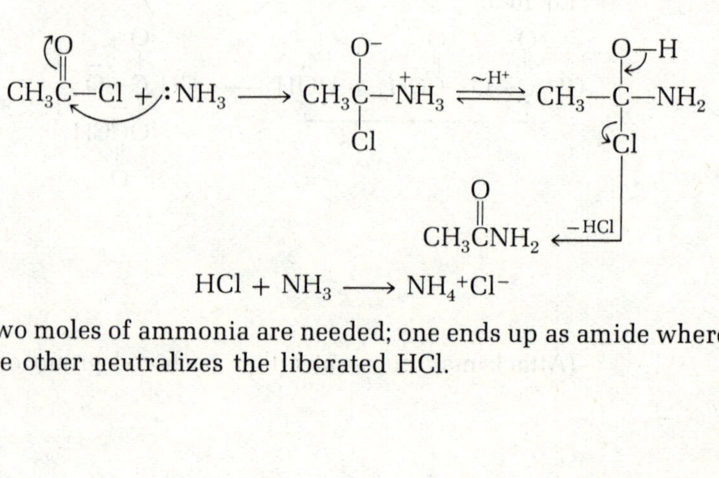

$$HCl + NH_3 \longrightarrow NH_4^+Cl^-$$

Two moles of ammonia are needed; one ends up as amide whereas the other neutralizes the liberated HCl.

Eq 10-41

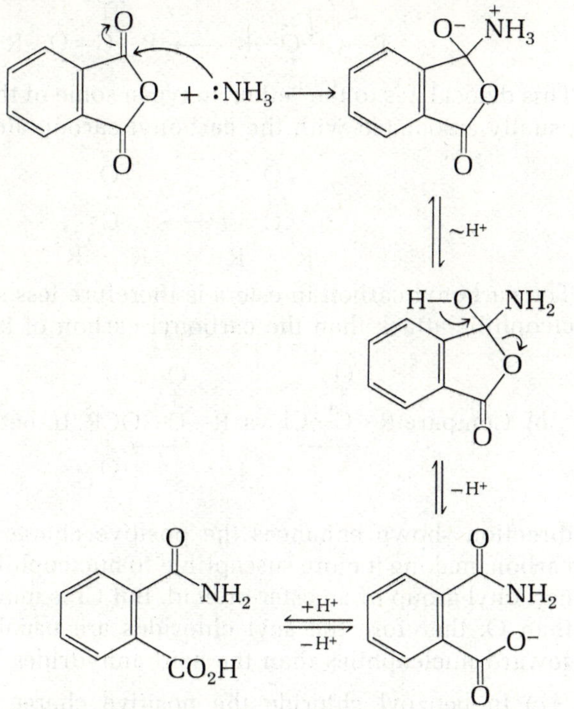

10-12.

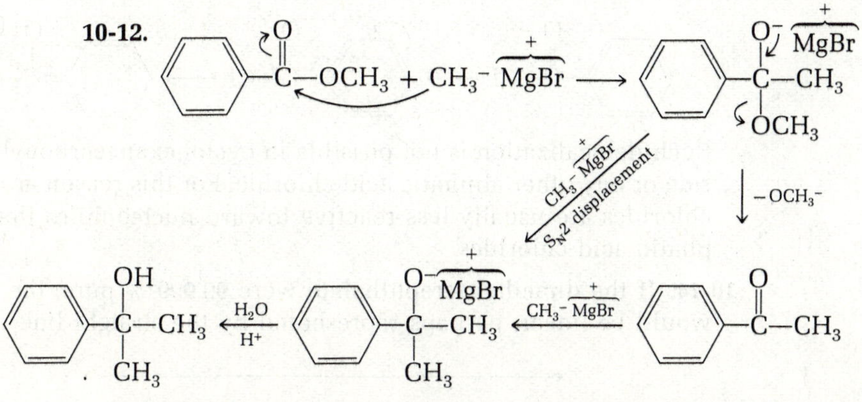

The reaction of esters with excess Grignard reagent provides a good method for making tertiary alcohols in which at least two of the three R groups attached to the alcohol carbon are identical.

10-13. a) The carbonyl group in esters is less reactive than the carbonyl group of ketones because of the resonance possibility shown on the next page:

$$R-\overset{\overset{\displaystyle O}{\|}}{C}-\overset{..}{\underset{..}{O}}-R \longleftrightarrow R-\overset{\overset{\displaystyle :\overset{..}{O}\overset{-}{}}{|}}{C}=\overset{+}{\underset{..}{O}}-R$$

This delocalizes to the "ether" oxygen some of the positive charge usually associated with the carbonyl carbon atom:

$$\underset{R\quad R}{\overset{\overset{\displaystyle O}{\|}}{C}} \longleftrightarrow \underset{R\quad R}{\overset{\overset{\displaystyle O^-}{|}}{C^+}}$$

The carbonyl carbon in esters is therefore less susceptible to nucleophilic attack than the carbonyl carbon of ketones.

b) Compare $R-\overset{\overset{\displaystyle O}{\|}}{C}-Cl$ vs. $R-\overset{\overset{\displaystyle O}{\|}}{C}-O\overset{\underset{\displaystyle \|}{\underset{\displaystyle O}{}}}{C}R$. In both, a dipole in the

direction shown enhances the positive charge on the carbonyl carbon, making it more susceptible to nucleophilic attack than the carbonyl group of an ester or acid. But Cl is more electronegative than O; therefore the acyl chlorides are usually more reactive toward nucleophiles than the acid anhydrides.

c) In benzoyl chloride the positive charge on the carbonyl carbon can be delocalized in the aromatic ring:

Such delocalization is not possible in cyclohexanecarbonyl chloride or any other aliphatic acid chloride. For this reason aryl acid chlorides are usually less reactive toward nucleophiles than aliphatic acid chlorides.

10-14. If the dimethyl terephthalate were 99.999+% pure, the chain would be linear, perhaps represented by the straight line

$$\xleftarrow{\hspace{3cm}\text{10 cm}\hspace{3cm}}\rightarrow$$

If the ester contained 2% of the meta isomer, then in 100 units there would be, on an average, two *meta* units. These would send the chain off twice at angles of 60°. This could be represented by the line

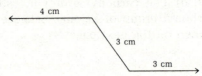

or

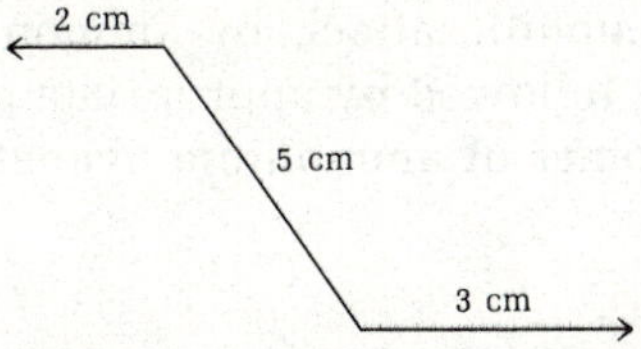

It is clear that the resulting polymer would have a *very* different shape from that of Dacron made with pure dimethyl terephthalate. The need for high-purity starting materials in polymerization reactions is critical in controlling the properties of the resulting polymers.

10-15. Since glycerol has three hydroxyl groups, the resulting ester can have chains which are "cross-linked," as well as branched at the middle or end hydroxyls.

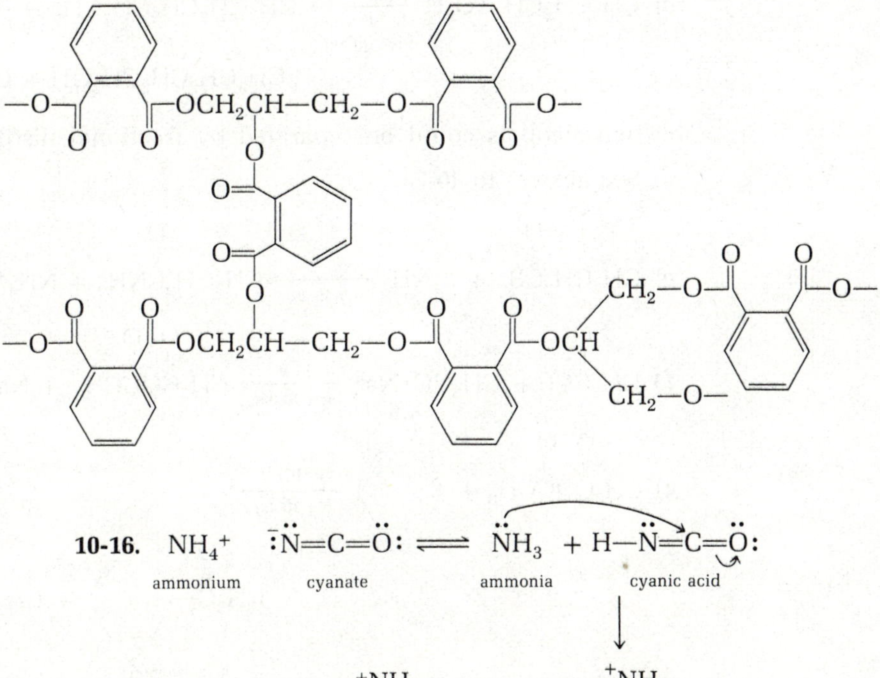

10-16. NH_4^+ $:N{=}C{=}\ddot{O}:$ $\rightleftharpoons$ $\ddot{N}H_3$ + $H{-}\ddot{N}{=}C{=}\ddot{O}:$

ammonium   cyanate   ammonia   cyanic acid

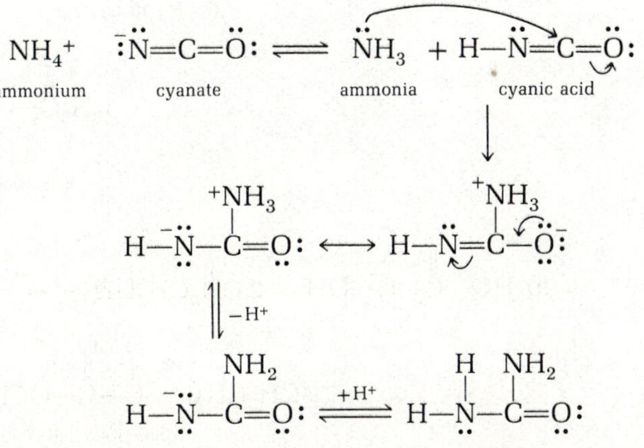

In the first step, proton transfer gives ammonia and cyanic acid.

111

Nucleophilic attack by ammonia on the carbon-oxygen double bond, followed by appropriate proton transfer then leads to urea, an isomer of ammonium cyanate.

10-17. *a)* $CH_3CH_2CH_2\overset{O}{\overset{\|}{C}}-Cl + CH_3OH \xrightarrow[\text{(eq 10-38)}]{}$

$$CH_3CH_2CH_2\overset{O}{\overset{\|}{C}}OCH_3 + HCl$$

b) $(CH_3CH_2\overset{O}{\overset{\|}{C}})_2O + 2\,NH_3 \xrightarrow[\text{(eq 10-41)}]{} 2\,CH_3CH_2\overset{O}{\overset{\|}{C}}-NH_2 + H_2O$

c) $CH_3CH_2CH_2CO_2H \xrightarrow[\substack{H^+ \\ \text{(eq 10-18)}}]{CH_3OH} CH_3CH_2CH_2CO_2CH_3 \xrightarrow[\text{(eq 10-26)}]{LiAlH_4}$

$$CH_3CH_2CH_2CH_2OH + CH_3OH$$

The two alcohols could be separated by fractional distillation.

d) See answer to 10-7a.

e) $CH_3CH_2\overset{O}{\overset{\|}{C}}Br + 2\,NH_3 \xrightarrow[\text{(eq 10-40)}]{} CH_3CH_2\overset{O}{\overset{\|}{C}}NH_2 + NH_4{}^+Br^-$

f) $CH_3\overset{O}{\overset{\|}{C}}Cl + CH_3\overset{O}{\overset{\|}{C}}O^-Na^+ \xrightarrow[\text{(eq 10-33)}]{\text{heat}} CH_3\overset{O}{\overset{\|}{C}}O\overset{O}{\overset{\|}{C}}CH_3 + Na^+Cl^-$

g) $CH_3\overset{O}{\overset{\|}{C}}O\overset{O}{\overset{\|}{C}}CH_3 + $ $\xrightarrow[\text{(eq 10-42)}]{AlCl_3}$

$$CH_3\overset{O}{\overset{\|}{C}}- \text{(benzene ring)} + CH_3CO_2H$$

acetophenone

h) $HO-\overset{O}{\overset{\|}{C}}-\overset{O}{\overset{\|}{C}}-OH + 2\,CH_3CH_2OH \xrightarrow{H^+}$

$$CH_3CH_2O-\overset{O}{\overset{\|}{C}}-\overset{O}{\overset{\|}{C}}-OCH_2CH_3 \qquad \text{(eq 10-18)}$$

diethyl oxalate

i) See equation 10-52.

j) [benzene ring]—COCl + [benzene ring] $\xrightarrow[\text{(eq 10-42)}]{\text{AlCl}_3}$

[diphenyl ketone structure]

+ HCl

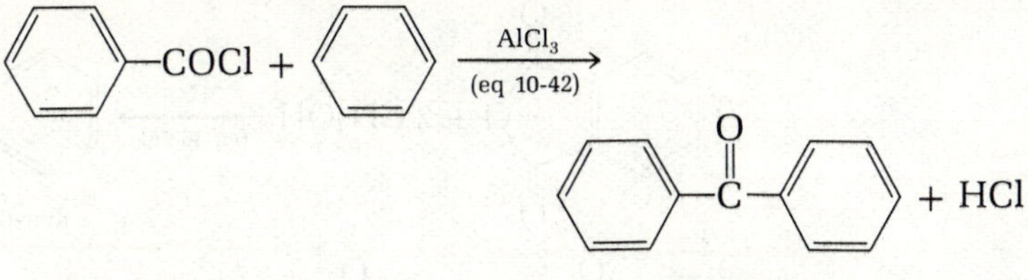

benzophenone

10-18. a) 2 [benzene ring]—CO$_2$H + HOCH$_2$CH$_2$OH $\xrightarrow{\text{H}^+}$

[structure: benzene ring—C(=O)—OCH$_2$CH$_2$O—C(=O)—benzene ring] + 2 H$_2$O   (eq 10-18)

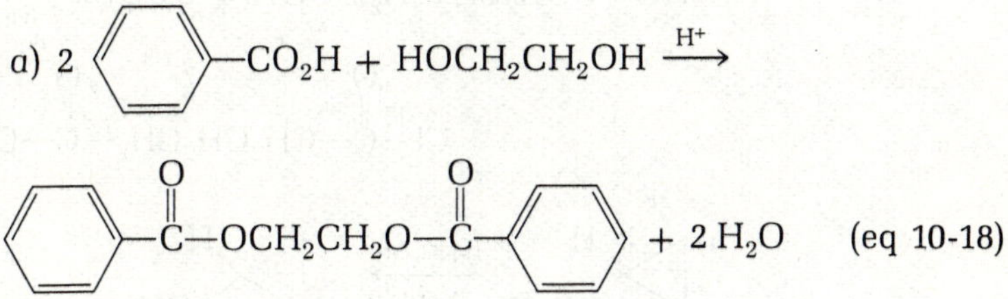

ethylene glycol dibenzoate

b) [benzene ring]—CH$_2$MgBr + O=C=O $\xrightarrow{\text{(eq 10-15)}}$

[structure: benzene ring—CH$_2$—C(=O)—O$^-$MgBr$^+$] $\xrightarrow{\text{H}_3\text{O}^+}$ [benzene ring]—CH$_2$CO$_2$H

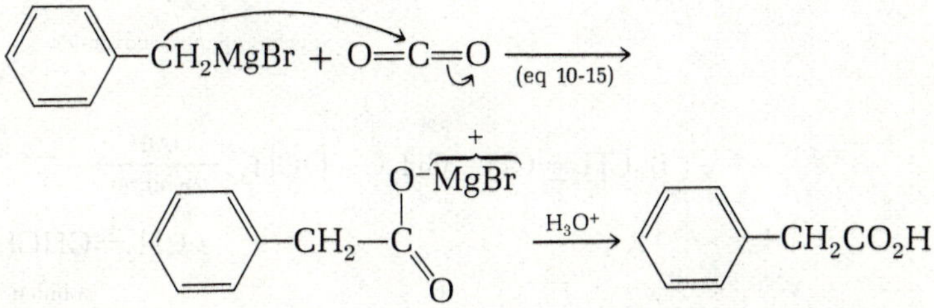

phenylacetic acid

c) O$_2$N—[benzene ring]—C(=O)—Cl + Na$^+$ $^-$OC(=O)CH$_3$ $\longrightarrow$

O$_2$N—[benzene ring]—C(=O)—O—C(=O)CH$_3$ + Na$^+$Cl$^-$ (eq 10-33)

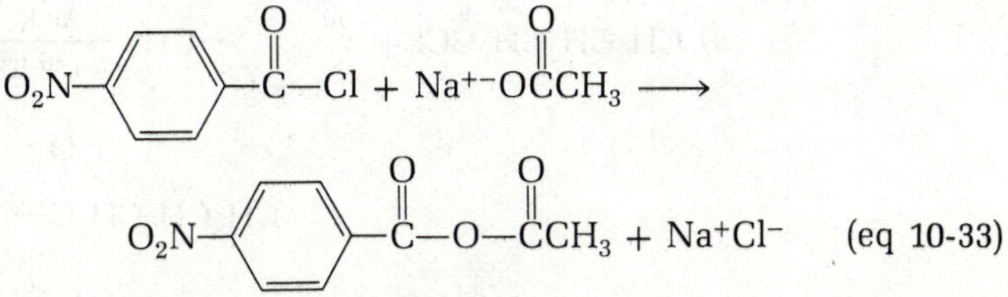

acetic p-nitrobenzoic anhydride

d) CH$_3$CH=CHCO$_2$H $\xrightarrow[\text{(eq 10-6)}]{\text{KMnO}_4 \atop \text{H}^+}$ CH$_3$CO$_2$H + [HO—C(=O)—C(=O)—OH]

acetic acid

2 CO$_2$ + H$_2$O $\xleftarrow[\text{H}^+]{\text{KMnO}_4}$

(sec 10.10)

e) [benzene ring]—CH$_2$CH$_2$CH$_3$ $\xrightarrow[\text{(eq 10-3)}]{\text{K}_2\text{Cr}_2\text{O}_7 \atop \text{H}^+}$ [benzene ring]—CO$_2$H

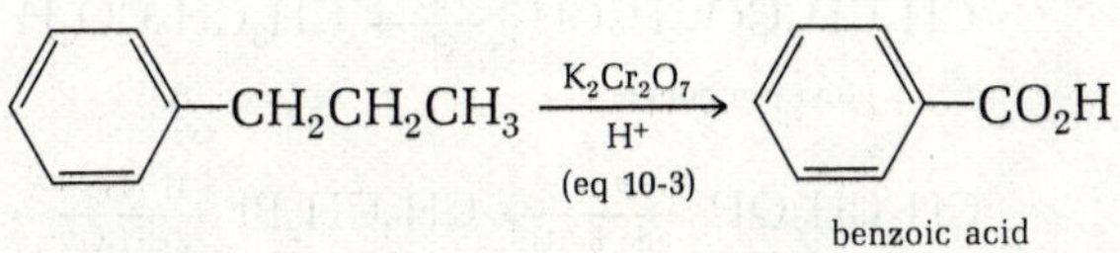

benzoic acid

113

f) $+ 2\ CH_3OH \xrightarrow[\text{(eq 10-18)}]{H^+}$ $+ H_2O$

dimethyl phthalate

g) $HO-\overset{O}{\underset{||}{C}}CH_2CH_2CH_2\overset{O}{\underset{||}{C}}-OH + 2\ SOCl_2 \xrightarrow[\text{(eq 10-31)}]{}$

$Cl-\overset{O}{\underset{||}{C}}-CH_2CH_2CH_2-\overset{O}{\underset{||}{C}}-Cl + 2\ HCl + 2\ SO_2$

h) $\xrightarrow[\text{2) heat}]{\text{1) NH}_4\text{OH}}$ (eq 10-46)

cyclopropanecarboxamide

i) $CH_2{=}CH-CH_2\overset{O}{\underset{||}{C}}-OCH_3 \xrightarrow[\text{(eq 10-26)}]{LiAlH_4}$

$CH_2{=}CHCH_2CH_2OH + CH_3OH$

3-buten-1-ol

j) $CH_3CH_2CH_2\overset{O}{\underset{||}{C}}Cl +$ $-CH_3 \xrightarrow[\text{(eq 10-42)}]{AlCl_3}$

$CH_3CH_2CH_2\overset{O}{\underset{||}{C}}-$ $-CH_3 + HCl$

n-propyl p-tolyl ketone

10-19. The original ester must have been derived from a 3-carbon acid **A** and a 2-carbon alcohol **B**, because the alcohol, when converted to the bromide **C** gave, through the cyanide route (which adds one carbon atom to the chain), the acid **A**. The only possible structure for the ester is ethyl propionate. The equations are:

$CH_3CH_2\overset{O}{\underset{||}{C}}OCH_2CH_3 \xrightarrow[H^+]{H_2O} CH_3CH_2CO_2H + CH_3CH_2OH$

ethyl propionate $\qquad\qquad$ **A** $\qquad\qquad$ **B**

$CH_3CH_2OH \xrightarrow[\text{(eq 6-31)}]{PBr_3} CH_3CH_2Br \xrightarrow[\text{2) H}_3\text{O}^+]{\text{1) NaCN}} CH_3CH_2CO_2H$

$\qquad\qquad\qquad\qquad\qquad$ **C** $\qquad$ (eq 10-10) $\qquad$ **A**

10-20. Since oxidation of an alcohol (B) to an acid occurs without a loss in carbon atoms, the original ester must have been derived from an alcohol and an acid with equal numbers of carbon atoms. The two possibilities are

$$\underset{\textbf{1}}{CH_3CH_2\overset{\displaystyle O}{\overset{\|}{C}}OCH_2CH_2CH_3} \quad \text{and} \quad \underset{\textbf{2}}{CH_3CH_2\overset{\displaystyle O}{\overset{\|}{C}}OCH(CH_3)_2}$$

However **2** can be discarded, because hydrolysis would give a secondary alcohol (isopropyl alcohol), which on oxidation would give a ketone (acetone; sec 9.4a) rather than an acid. Therefore the original ester must have been *n*-propyl propionate **(1)**, and the equations are:

$$CH_3CH_2\overset{\displaystyle O}{\overset{\|}{C}}OCH_2CH_2CH_3 \xrightarrow[H^+]{H_2O} \underset{A}{CH_3CH_2CO_2H} + \underset{B}{CH_3CH_2CH_2OH}$$

$$\underset{B}{CH_3CH_2CH_2OH} \xrightarrow[(eq\ 10\text{-}7)]{[O]} \underset{A}{CH_3CH_2CO_2H}$$

CHAPTER ELEVEN

FATS, OILS, WAXES, AND DETERGENTS

11-1. Use Table 11-1 as a nomenclature guide.

a) $CH_3(CH_2)_{14}CO_2^-K^+$

b) $[CH_3(CH_2)_7CH{=}CH(CH_2)_7CO_2^-]_2Mg^{2+}$

c) $CH_3(CH_2)_{12}CH_2OSO_3^-Na^+$

d)
$$CH_2{-}O{-}\overset{\displaystyle O}{\overset{\|}{C}}(CH_2)_{10}CH_3$$
$$|$$
$$CH{-}O{-}\overset{\displaystyle O}{\overset{\|}{C}}(CH_2)_{10}CH_3$$
$$|$$
$$CH_2{-}O{-}\overset{\displaystyle O}{\overset{\|}{C}}(CH_2)_{10}CH_3$$

e)

$$CH_2-O-\overset{\overset{\displaystyle O}{\|}}{C}CH_2CH_2CH_3$$

$$CH-O-\overset{\overset{\displaystyle O}{\|}}{C}(CH_2)_{14}CH_3$$

$$CH_2-O-\overset{\overset{\displaystyle O}{\|}}{C}(CH_2)_7CH=CH(CH_2)_7CH_3$$

f)

$$CH_2-O-\overset{\overset{\displaystyle O}{\|}}{C}(CH_2)_{16}CH_3$$

$$CH-O-\overset{\overset{\displaystyle O}{\|}}{C}(CH_2)_{14}CH_3$$

$$CH_2-O-\overset{\overset{\displaystyle O}{\|}}{C}(CH_2)_{16}CH_3$$

g) $CH_3(CH_2)_4CH=CHCH_2CH=CH(CH_2)_7\overset{\overset{\displaystyle O}{\|}}{C}-O-(CH_2)_{13}CH_3$

h) $C_8H_{17}-\langle\!\!\!\bigcirc\!\!\!\rangle-SO_3^-Na^+$

11-2.

Saponification

$$CH_2-O-\overset{\overset{\displaystyle O}{\|}}{C}(CH_2)_7CH=CHCH_2CH=CHCH_2CH=CHCH_2CH_3$$

$$CH-O-\overset{\overset{\displaystyle O}{\|}}{C}(CH_2)_7CH=CHCH_2CH=CHCH_2CH=CHCH_2CH_3 \;+\; 3\,NaOH \xrightarrow{\text{(eq 11-2)}}$$

$$CH_2-O-\overset{\overset{\displaystyle O}{\|}}{C}(CH_2)_7CH=CHCH_2CH=CHCH_2CH=CHCH_2CH_3$$

$$\begin{array}{l} CH_2OH \\ CHOH \;+\; 3\,CH_3CH_2CH=CHCH_2CH=CHCH_2CH=CH(CH_2)_7CO_2^-Na^+ \\ CH_2OH \end{array}$$

Hydrogenation

$$CH_2-O-\overset{\overset{\displaystyle O}{\|}}{C}(CH_2)_7CH=CHCH_2CH=CHCH_2CH=CHCH_2CH_3 \qquad\qquad CH_2-O-\overset{\overset{\displaystyle O}{\|}}{C}(CH_2)_{16}CH_3$$

$$CH-O-\overset{\overset{\displaystyle O}{\|}}{C}(CH_2)_7CH=CHCH_2CH=CHCH_2CH=CHCH_2CH_3 \;+\; 9\,H_2 \xrightarrow[\text{Heat}]{\text{Ni}} CH-O-\overset{\overset{\displaystyle O}{\|}}{C}(CH_2)_{16}CH_3$$

$$CH_2-O-\overset{\overset{\displaystyle O}{\|}}{C}(CH_2)_7CH=CHCH_2CH=CHCH_2CH=CHCH_2CH_3 \qquad\qquad CH_2-O-\overset{\overset{\displaystyle O}{\|}}{C}(CH_2)_{16}CH_3$$

(eq 11-3)

Hydrogenolysis

$$\begin{array}{l} CH_2-O-\overset{\overset{O}{\|}}{C}(CH_2)_7CH=CHCH_2CH=CHCH_2CH=CHCH_2CH_3 \\[2mm] CH-O-\overset{\overset{O}{\|}}{C}(CH_2)_7CH=CHCH_2CH=CHCH_2CH=CHCH_2CH_3 \quad + 6\,H_2 \xrightarrow[\text{chromite}]{\text{zinc}} \\[2mm] CH_2-O-\overset{\overset{O}{\|}}{C}(CH_2)_7CH=CHCH_2CH=CHCH_2CH=CHCH_2CH_3 \end{array}$$

(eq 11-4)

$$\begin{array}{l} CH_2OH \\ CHOH \quad + \ 3\,CH_3CH_2CH=CHCH_2CH=CHCH_2CH=CH(CH_2)_7CH_2OH \\ CH_2OH \end{array}$$

11-3. *a*) glyceryl tripalmitate (palmitin)

   b) β-stearo-α,α'-dimyristin

   c) myristyl linoleate

   d) glyceryl oleylstearyllaurate

11-4. The essentials of soap and detergent design are discussed in sections 11.3b and 11.3d.

11-5. Fruit juices may be acidic due to the presence of citric and other acids. Ordinary soaps are not very effective in acidic water because of the reaction shown in equation 11-5.

11-6. The calcium and magnesium ions present in hard water form insoluble salts with ordinary soaps (eq 11-6). In contrast, the calcium and magnesium salts of sulfate or sulfonate detergents (sec 11.3d) are water soluble. Nonacidic syndets such as the esters of polyols or the tetraalkyl ammonium ions are also not affected by these metallic ions.

11-7. Catalytic hydrogenation of each acid would give the same product, stearic acid, thus proving that all three had a linear chain of 18 carbon atoms.

11-8. Ozonolysis would lead to the following products:

$$CH_3(CH_2)_7CH=CH(CH_2)_7CO_2H \xrightarrow[\text{(eq 3-26)}]{}$$
oleic acid

$$CH_3(CH_2)_7CH=O \ + \ O=CH(CH_2)_7CO_2H$$

$$CH_3(CH_2)_4CH=CHCH_2CH=CH(CH_2)_7CO_2H \longrightarrow$$
linoleic acid

$$CH_3(CH_2)_4CH=O \ + \ O=CHCH_2CH=O \ + \ O=CH(CH_2)_7CO_2H$$

$$CH_3CH_2CH=CHCH_2CH=CHCH_2CH=CH(CH_2)_7CO_2H \longrightarrow$$
linolenic acid

$$CH_3CH_2CH=O \ + \ 2\,O=CHCH_2CH=O \ + \ O=CH(CH_2)_7CO_2H$$

Identification of the different products would lead in each case to an unambiguous assignment of the double bond positions (but not their stereochemistry).

11-9. For the synthesis and structure of propylene tetramer, see the answer to problem 3-20.

$$CH_3CHCH_2CHCH_2CHCH_2CH{=}CH_2 \; + \; \underset{\text{(eq 5-25)}}{\bigcirc} \xrightarrow[\substack{\text{or}\\H^+}]{AlCl_3}$$
$$\underset{CH_3}{|} \quad \underset{CH_3}{|} \quad \underset{CH_3}{|}$$

$$CH_3CHCH_2CHCH_2CHCH_2CH{-}\bigcirc \xrightarrow[\text{(eq 5-8)}]{H_2SO_4}$$
$$\underset{CH_3}{|} \quad \underset{CH_3}{|} \quad \underset{CH_3}{|} \quad \underset{CH_3}{|}$$

$$CH_3CHCH_2CHCH_2CHCH_2CH{-}\bigcirc{-}SO_3H \xrightarrow[\text{(eq 11-8)}]{Na^+OH^-}$$
$$\underset{CH_3}{|} \quad \underset{CH_3}{|} \quad \underset{CH_3}{|} \quad \underset{CH_3}{|}$$

$$CH_3CHCH_2CHCH_2CHCH_2CH{-}\bigcirc{-}SO_3^-Na^+$$
$$\underset{CH_3}{|} \quad \underset{CH_3}{|} \quad \underset{CH_3}{|} \quad \underset{CH_3}{|}$$

The detergent produced this way was, for many years, one of the leading syndets. The branched-chain structure, however, cannot be readily degraded biologically according to the scheme given in section 11.7, because branching prevents stepwise two-carbon acetate formation.

11-10. $CH_3(CH_2)_7CH{=}CH_2 \; + \; \underset{\text{(eq 5-25)}}{\bigcirc} \xrightarrow{AlCl_3}$

$$CH_3(CH_2)_7CH{-}\bigcirc \xrightarrow[\text{(eq 5-8)}]{H_2SO_4}$$
$$\underset{CH_3}{|}$$

$$CH_3(CH_2)_7CH{-}\bigcirc{-}SO_3H \xrightarrow[\text{(eq 11-8)}]{Na^+OH^-}$$
$$\underset{CH_3}{|}$$

$$CH_3(CH_2)_7CH{-}\bigcirc{-}SO_3^-Na^+$$
$$\underset{CH_3}{|}$$

Biological degradation begins at the end of the alkyl chain (i.e., at the terminal methyl group). The normal fatty acid degradation (eq 11-10 and 11-11) can occur down to the carbon α to the aromatic ring.

11-11. The reactions are analogous to equation 7-21.

$$CH_3(CH_2)_{14}CH_2OH + CH_2\text{—}CH_2 \longrightarrow$$
$$\underset{O}{\diagdown\diagup}$$

$$CH_3(CH_2)_{14}CH_2OCH_2CH_2OH$$

$$CH_3(CH_2)_{14}CH_2OCH_2CH_2OH + CH_2\text{—}CH_2 \longrightarrow$$
$$\underset{O}{\diagdown\diagup}$$

$$CH_3(CH_2)_{14}CH_2[OCH_2CH_2]_2OH$$

The reaction is performed using sufficient ethylene oxide to build the desired range of ether groups (8–12) in the molecule.

11-12. Drying oils are fats derived from highly unsaturated acids; put another way, they are derived from fats which have a high iodine number. Drying oils are used in oil-based paints, oilcloth, and linoleum.

11-13. Hydrogenation of an oil destroys its drying properties by removing the double bonds essential to the polymerization and oxidation reactions which constitute "drying." Conversely, any method which increases the number of double bonds in an oil enhances its drying properties; for example, addition and elimination of halogen would accomplish this:

$$-CH_2\text{—}CH=CH\text{—}CH_2- \xrightarrow{X_2} -CH_2\underset{X}{C}H\underset{X}{C}HCH_2- \xrightarrow{\text{base}}$$

$$-CH=CH\text{—}CH=CH-$$

11-14. Saponification number is defined as the number of milligrams of KOH required to hydrolyze 1 g of a fat. Since 3 moles of KOH are required for every mole of fat (eq 11-1), $3 \times 56 = 168$ g of KOH are required per mole of fat. Therefore, (1/Mol. Wt. of fat) $\times$ 168 g of KOH will be needed to hydrolyze 1 g of fat. To convert to milligrams of KOH, one must then multiply by 1000, so the correct formula is

$$\text{Saponification No.} = \frac{168000}{\text{Mol. Wt. of fat}}$$

a) $\dfrac{168,000}{302} = 556$

b) $\dfrac{168,000}{806} = 208$

11-15. The formula developed in answer to problem 11-14 can be used.

$$190 = \frac{168{,}000}{\text{Mol. Wt.}}$$

$$\text{Mol. Wt.} = \frac{168{,}000}{190} = 884$$

11-16. Iodine number is defined as the number of grams of iodine that will combine with 100 g of a fat.

a) Glyceryl trioleate has a molecular weight of 884; it will react with 3 moles of I_2, since it contains 3 double bonds. The atomic weight of iodine is 127. Therefore, 884 g of glyceryl trioleate will react with $3 \times 254 = 762$ g of iodine. Therefore, 100 g of the fat will react with $\frac{100}{884} \times 762 = 86.2$ g of iodine. Therefore, the *iodine number* of glyceryl trioleate is 86.2.

b) The reasoning is as described in part *a*. The molecular weight of the fat is 872; it will react with 9 moles of iodine. Therefore, the iodine number of this glyceride is $\frac{100}{872} \times 2286 = 262.1$.

11-17. One hundred grams of the fat reacts with 95.3 g of iodine. Therefore, one mole of the fat will react with $8 \times 95.3 = 762.4$ g of iodine. If the fat had one double bond per molecule, then one mole would react with one mole or 254 g of iodine. Therefore, the average number of double bonds per fat molecule is $\frac{762.4}{254} = 3$.

11-18. Compare the generalized formulas for a fat or oil—a glyceride, equation 11-1—with that of a wax—a simple ester, section 11.6.

$$\overset{\displaystyle O}{\overset{\displaystyle \|}{C_{15}H_{31}C}}\!-\!OC_{16}H_{33} + Na^+OH^- \longrightarrow C_{15}H_{31}CO_2^-Na^+ + C_{16}H_{33}OH$$

cetyl palmitate

11-19. *a)* $CH_3CH_2CH_2CO_2H + CoA\!-\!SH \xrightarrow[\text{(eq 11-9)}]{\text{enzyme}}$

$$\overset{\displaystyle O}{\overset{\displaystyle \|}{CH_3CH_2CH_2C}}\!-\!S\!-\!CoA + H_2O$$

$$\overset{\displaystyle O}{\overset{\displaystyle \|}{CH_3CH_2CH_2C}}\!-\!S\!-\!CoA + \overset{\displaystyle O \quad\; O}{\overset{\displaystyle \| \quad\; \|}{HO\!-\!C\!-\!CH_2C}}\!-\!S\!-\!CoA \underset{\text{(eq 11-12)}}{\overset{\text{enzyme}}{\rightleftharpoons}}$$

malonyl coenzyme A

$$\overset{\displaystyle O \qquad\;\; O}{\overset{\displaystyle \| \qquad\;\; \|}{HO\!-\!C\!-\!CH\!-\!C}}\!-\!S\!-\!CoA + CoA\!-\!SH \underset{\text{enzyme}}{\overset{-CO_2}{\rightleftharpoons}}$$

$$\underset{CH_3CH_2CH_2C=O}{|} \qquad\qquad\qquad \text{(eq 11-12)}$$

$$\overset{\displaystyle O \quad\; O}{\overset{\displaystyle \| \quad\; \|}{CH_3CH_2CH_2CCH_2C}}\!-\!S\!-\!CoA$$

$$CH_3CH_2CH_2\overset{O}{\underset{\|}{C}}CH_2\overset{O}{\underset{\|}{C}}-S-CoA \underset{\substack{\text{enzyme} \\ \text{(eq 11-10)}}}{\overset{+2H}{\rightleftharpoons}} CH_3CH_2CH_2\overset{OH}{\underset{|}{C}HCH_2\overset{O}{\underset{\|}{C}}-S-CoA}$$

$$\Big\updownarrow \substack{-H_2O \quad \text{enzyme}}$$

$$CH_3CH_2CH_2CH_2CH_2\overset{O}{\underset{\|}{C}}-S-CoA \underset{\substack{\text{enzyme} \\ \text{(eq 11-10)}}}{\overset{+2H}{\rightleftharpoons}} CH_3CH_2CH_2CH=CH\overset{O}{\underset{\|}{C}}-S-CoA$$

$$CH_3CH_2CH_2CH_2CH_2\overset{O}{\underset{\|}{C}}-S-CoA + H_2O \underset{\text{(eq 11-9)}}{\overset{\text{enzyme}}{\rightleftharpoons}}$$

$$CH_3CH_2CH_2CH_2CH_2CO_2H + CoA-SH$$

b) $$CH_3CH_2CH_2CH_2CH_2\overset{O}{\underset{\|}{C}}-OH + CoA-SH \underset{\text{(eq 11-9)}}{\overset{\text{enzyme}}{\rightleftharpoons}}$$

$$CH_3CH_2CH_2CH_2CH_2\overset{O}{\underset{\|}{C}}-S-CoA + H_2O$$

$$CH_3(CH_2)_4\overset{O}{\underset{\|}{C}}-S-CoA \underset{\substack{\text{enzyme} \\ \text{(eq 11-10)}}}{\overset{-2H}{\rightleftharpoons}} CH_3CH_2CH_2CH=CH-\overset{O}{\underset{\|}{C}}-S-CoA$$

$$\Big\updownarrow \substack{+H_2O \quad \text{enzyme} \\ \text{(eq 11-10)}}$$

$$CH_3CH_2CH_2\overset{O}{\underset{\|}{C}}CH_2\overset{O}{\underset{\|}{C}}-S-CoA \underset{\substack{\text{enzyme} \\ \text{(eq 11-10)}}}{\overset{-2H}{\rightleftharpoons}} CH_3CH_2CH_2\overset{OH}{\underset{|}{C}}HCH_2\overset{O}{\underset{\|}{C}}-S-CoA$$

$$CH_3CH_2CH_2\overset{O}{\underset{\|}{C}}CH_2\overset{O}{\underset{\|}{C}}-S-CoA + CoA-SH \underset{\text{(eq 11-11)}}{\overset{\text{enzyme}}{\rightleftharpoons}}$$

$$CH_3CH_2CH_2\overset{O}{\underset{\|}{C}}-S-CoA + CH_3\overset{O}{\underset{\|}{C}}-S-CoA$$

$$CH_3CH_2CH_2\overset{O}{\underset{\|}{C}}-S-CoA + H_2O \underset{\text{(eq 11-9)}}{\overset{\text{enzyme}}{\rightleftharpoons}}$$

$$CH_3CH_2CH_2CO_2H + CoA-SH$$

11-20. Since the fat is built up two carbon atoms at a time from the carboxyl end, one would expect alternate carbons to be labeled. These are shown with asterisks in the formula

$$\overset{*}{C}H_3CH_2\overset{*}{C}H_2CH_2\overset{*}{C}H_2CH_2\overset{*}{C}H_2CH_2\overset{*}{C}H_2CH_2\overset{*}{C}H_2CH_2\overset{*}{C}H_2CH_2\overset{*}{C}H_2CO_2H$$

To verify this, start with CH_3COS—CoA in equation 11-12, and build up to butyric acid. The pattern then repeats itself.

CHAPTER TWELVE

AMINES AND DIAZONIUM COMPOUNDS

12-1. Many correct answers are possible; only one example will be given in each case.

a) CH_3NH_2 methylamine

b)   pyrrolidine

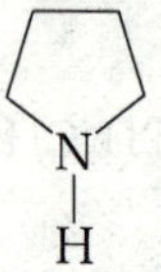

c)   N,N-dimethylaniline

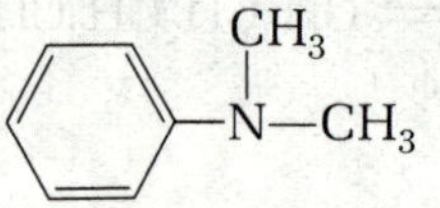

d)   tetramethylammonium chloride

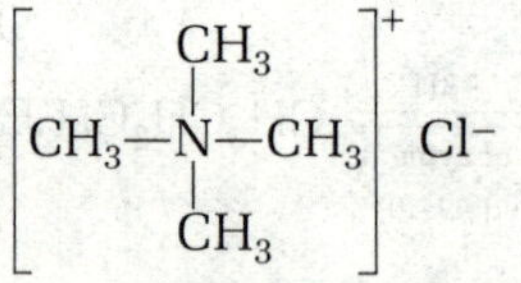

e)   benzenediazonium chloride

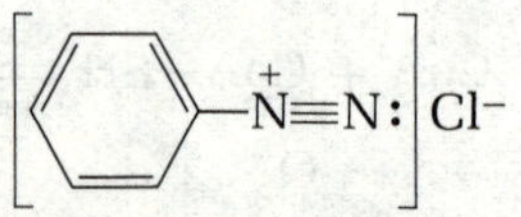

f)   pyridine

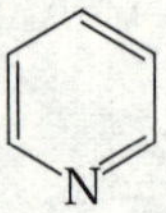

g)   azobenzene

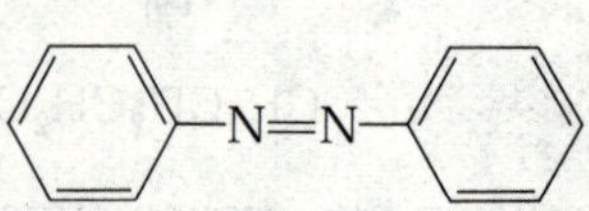

h)   N-nitrosodimethylamine

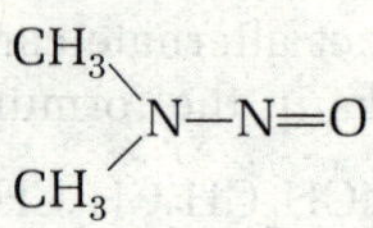

12-2. a) 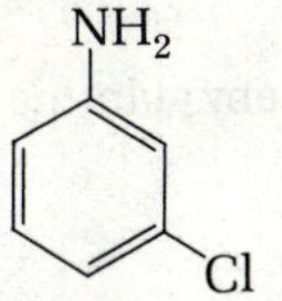 NH_2

Cl

b) $CH_3CHCH_2CH_3$
       NH_2

c) $CH_3CHCH_2CH_2CH_2CH_3$
       NH_2

d) $CH_3CH_2CH_2NCH_3$
           CH_3

e) 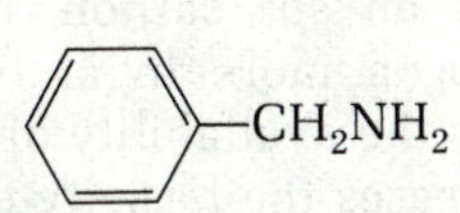 ⬡—CH_2NH_2

f) $CH_3CH—CH_2$
       NH_2 NH_2

g) 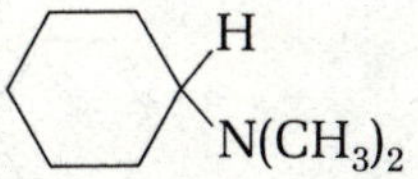 ⬡$\begin{smallmatrix}H\\ \diagdown N(CH_3)_2\end{smallmatrix}$

h) $(CH_3CH_2)_4N^+Br^-$

i) 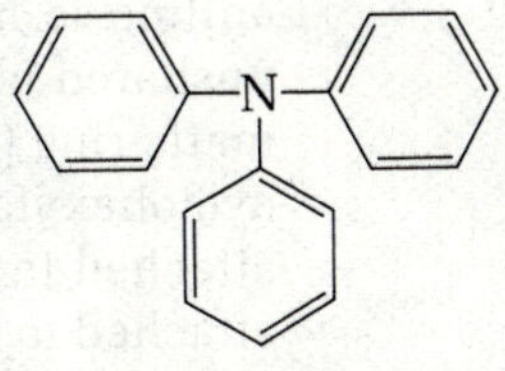 ⬡—N(⬡)⬡

j) 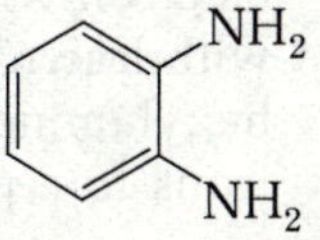 ⬡$\begin{smallmatrix}NH_2\\ \\ NH_2\end{smallmatrix}$

12-3. a) p-bromoaniline
b) methylisopropylamine
c) methyldiethylamine
d) tetramethylammonium chloride
e) 4-amino-2-butanol
f) 2-methylaminoethanol
g) benzenediazonium hydrogen sulfate
h) 2-aminonaphthalene
i) cyclopentylamine
j) 1,3-diaminopropane

12-4. a) $CH_3CH_2CH_2CH_2NH_2$ n-butylamine (primary)
     $CH_3CH_2CHCH_3$ *sec*-butylamine (primary)
            NH_2
     $CH_3CHCH_2NH_2$ isobutylamine (primary)
        CH_3
     $(CH_3)_3CNH_2$ t-butylamine (primary)
     $CH_3CH_2CH_2NHCH_3$ methyl-n-propylamine (secondary)
     $CH_3CHNHCH_3$ methyl-i-propylamine (secondary)
        CH_3
     $CH_3CH_2NHCH_2CH_3$ diethylamine (secondary)
     $CH_3—N—CH_2CH_3$ dimethylethylamine (tertiary)
         CH_3

b) 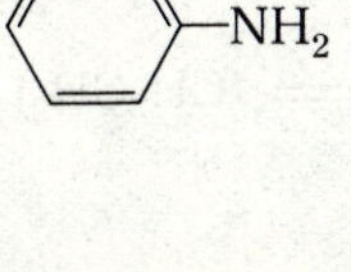 CH_3

⬡—NH_2 o-toluidine (primary)
                  and the m- and p-isomers

$\text{C}_6\text{H}_5\text{—CH}_2\text{NH}_2$ benzylamine (primary)

$\text{C}_6\text{H}_5\text{—NHCH}_3$ N-methylaniline (secondary)

12-5. *a*) The question is perhaps best answered by explaining why aniline is a weaker base than cyclohexylamine. In aniline, the unshared electron pair on the nitrogen is delocalized to the aromatic ring (sec 12.3). No analogous delocalization is possible with cyclohexylamine. Second, in aniline an sp^2 hybridized carbon is attached to the nitrogen, whereas in cyclohexylamine the carbon attached to nitrogen is sp^3 hybridized. Since an sp^2 carbon is more electron-withdrawing than an sp^3 carbon, electron density is withdrawn from the nitrogen more in aniline than in cyclohexylamine. This decreases the availability of the unshared electrons to a proton; i.e., decreases the basicity of the nitrogen.

b) The nitro group is electron-withdrawing; electron density is thus withdrawn from the amino nitrogen, making it less basic toward a proton.

12-6. Hydrogen bonding is possible in dimethylamine, but not in trimethylamine:

$$(\text{CH}_3)_2\text{N—H} \text{-----} \overset{\displaystyle \text{CH}_3}{\underset{\displaystyle \text{CH}_3}{\text{N}}}\text{—H---}\qquad \text{etc.}$$

Energy is required to break these hydrogen bonds, to vaporize dimethylamine.

12-7. $\text{H}_3\text{N:} + \text{CH}_3\text{CH}_2\text{—Cl} \longrightarrow \overset{+}{\text{H}_3\text{N}}\text{CH}_2\text{CH}_3 \;\; \text{Cl}^-$

$\overset{+}{\text{H}_3\text{N}}\text{CH}_2\text{CH}_3 \; \text{Cl}^- + \text{NH}_3 \rightleftharpoons \text{H}_2\text{NCH}_2\text{CH}_3 + \text{NH}_4{}^+\text{Cl}^-$

$\text{CH}_3\text{CH}_2\ddot{\text{N}}\text{H}_2 + \text{CH}_3\text{CH}_2\text{—Cl} \longrightarrow (\text{CH}_3\text{CH}_2)_2\overset{+}{\text{N}}\text{H}_2 \;\; \text{Cl}^-$

$(\text{CH}_3\text{CH}_2)_2\overset{+}{\text{N}}\text{H}_2 \; \text{Cl}^- + \text{NH}_3 \rightleftharpoons (\text{CH}_3\text{CH}_2)_2\text{NH} + \text{NH}_4{}^+\text{Cl}^-$

$(\text{CH}_3\text{CH}_2)_2\ddot{\text{N}}\text{H} + \text{CH}_3\text{CH}_2\text{—Cl} \longrightarrow (\text{CH}_3\text{CH}_2)_3\overset{+}{\text{N}}\text{H} \;\; \text{Cl}^-$

$(\text{CH}_3\text{CH}_2)_3\overset{+}{\text{N}}\text{H} \; \text{Cl}^- + \text{NH}_3 \rightleftharpoons (\text{CH}_3\text{CH}_2)_3\text{N} + \text{NH}_4{}^+\text{Cl}^-$

Monoalkylation can be favored by keeping the ratio of ethyl chloride to ammonia low.

12-8. *a*) One could start with a three-carbon alcohol and add the fourth carbon via the nitrile:

$$CH_3CH_2CH_2OH \xrightarrow[\text{(eq 6-31)}]{PBr_3} CH_3CH_2CH_2Br \xrightarrow[\text{(eq 8-11)}]{NaCN}$$

$$CH_3CH_2CH_2CN \xrightarrow[\text{Ni}]{H_2} CH_3CH_2CH_2CH_2NH_2$$
(eq 12-9)

Other plausible routes can be devised.

b) CH$_3$—⬡ $\xrightarrow[\text{(eq 5-8)}]{HONO_2}$ CH$_3$—⬡—NO$_2$ $\xrightarrow[\text{HCl}]{Sn}$
(eq 12-11)

CH$_3$—⬡—NH$_2$

c) Repeat the method used in part *a*, but start with 1-butanol.

d) ⬡ $\xrightarrow[\text{(eq 5-8, 12-11)}]{\substack{1.\ HONO_2 \\ 2.\ H_2/Pt}}$ ⬡—NH$_2$ $\xrightarrow[\text{(eq 12-8)}]{CH_3CH_2Br}$

⬡—NHCH$_2$CH$_3$

e) ⬡—CH$_3$ $\xrightarrow[\text{(eq 5-20)}]{KMnO_4}$ ⬡—CO$_2$H $\xrightarrow[\text{H}^+]{HONO_2}$
(eq 5-8)

O$_2$N—⬡—CO$_2$H $\xrightarrow[\text{(eq 12-11)}]{H_2/Pt}$ H$_2$N—⬡—CO$_2$H

The carboxyl group is *meta*-directing; thus it is better to nitrate the benzoic acid rather than the toluene, which would give mainly the *ortho* and *para* isomers.

f) CH$_3$CH$_2$CH$_2$CH$_2$OH $\xrightarrow[\text{(eq 6-31)}]{PBr_3}$ CH$_3$CH$_2$CH$_2$CH$_2$Br $\xrightarrow[\text{(eq 12-4)}]{NH_3}$

(CH$_3$CH$_2$CH$_2$CH$_2$)$_3$N

In the second step, use an excess of *n*-butyl bromide.

g) $CH_3CH_2OH \xrightarrow[\text{heat}]{H^+} CH_2{=}CH_2 \xrightarrow[\text{(eq 3-2)}]{Br_2} \underset{\overset{|}{Br} \quad \overset{|}{Br}}{CH_2{-}CH_2} \xrightarrow[\text{(eq 8-11)}]{2Na^+CN^-}$

(eq 3-29)

$\underset{\overset{|}{CN} \quad \overset{|}{CN}}{CH_2{-}CH_2} \xrightarrow[\text{Pt}]{4H_2} H_2NCH_2CH_2CH_2CH_2NH_2$

(eq 12-9)

h) $CH_3CH_2CH_2OH \xrightarrow[\text{(eq 10-7)}]{K_2Cr_2O_7} CH_3CH_2CO_2H \xrightarrow[\text{(eq 10-31)}]{SOCl_2}$

$CH_3CH_2COCl \xrightarrow[\text{(eq 10-40)}]{NH_3} CH_3CH_2\overset{\overset{\displaystyle O}{\|}}{C}NH_2 \xrightarrow[\text{(eq 12-12)}]{Br_2,\ NaOH} CH_3CH_2NH_2$

$CH_3CH_2\overset{\overset{\displaystyle O}{\|}}{C}Cl + CH_3CH_2NH_2 \xrightarrow[\text{(eq 12-26)}]{} CH_3CH_2\overset{\overset{\displaystyle O}{\|}}{C}NHCH_2CH_3$

$CH_3CH_2CH_2NHCH_2CH_3 \xleftarrow[\text{(eq 12-26)}]{LiAlH_4}$

12-9. a) The carbon chain length must be decreased; this suggests the Hofmann rearrangement.

$CH_3CH_2CH_2CONH_2 \xrightarrow[\text{NaOH}]{Br_2} CH_3CH_2CH_2NH_2$

(eq 12-12)

b) $CH_3CH_2CH_2CH{=}O + NH_2OH \xrightarrow[\text{(eq 9-29)}]{}$

$CH_3CH_2CH_2CH{=}NOH \xrightarrow[\text{Ni (eq 12-10)}]{2H_2} CH_3CH_2CH_2CH_2NH_2$

c) $-CH_2Br + Na^+CN^- \xrightarrow[\text{(eq 8-11)}]{}$

$-CH_2CN \xrightarrow[\text{Ni}]{2H_2} $$-CH_2CH_2NH_2$

(eq 12-9)

d) $-NO_2 \xrightarrow[\text{Fe Cl}_3]{Cl_2}$$-NO_2 \xrightarrow[\text{Pt}]{H_2}$$-NH_2$

(eq 5-8) (eq 12-11)

e) $CH_3CH_2\overset{\overset{\displaystyle O}{\|}}{C}CH_2CH_3 + NH_2OH \xrightarrow[\text{(eq 9-29)}]{} CH_3CH_2\overset{\overset{\displaystyle NOH}{\|}}{C}CH_2CH_3$

$\underset{\overset{|}{NH_2}}{CH_3CH_2CHCH_2CH_3} \xleftarrow[\text{Ni}]{2H_2}$

(eq 12-10)

12-10. $R-\ddot{N}=C=O + H-\ddot{O}-H \longrightarrow R-\ddot{N}=C-O^-$

$$R-\ddot{N}=C=O + H-\ddot{O}-H \longrightarrow R-\ddot{N}=C-O^-$$

with $\overset{+}{O}H_2$ (H, H)

$$\xrightarrow[\text{(eq 10-14)}]{\text{tautomerize}} \left[R-\ddot{N}=C \overset{OH}{\underset{OH}{\big<}} \right] \underset{+H^+}{\overset{-H^+}{\longleftrightarrow}}$$

$$\left[R-\underset{H}{\overset{O}{\underset{|}{\overset{\|}{N}}}}-C-O-H \right] \xrightarrow[+H^+]{-H^+} RNH_2 + CO_2$$

a carbamic acid

12-11. *a)* ⬡$-NH_2 + HCl \longrightarrow$ ⬡$-\overset{+}{N}H_3 + Cl^-$ (eq 12-19)

b) 2 ⬡$-NH_2 + CH_3\overset{O}{\overset{\|}{C}}Cl \longrightarrow$

⬡$-NH\overset{O}{\overset{\|}{C}}CH_3 + $ ⬡$-\overset{+}{N}H_3Cl^-$ (eq 10-40, 12-26)

c) ⬡$-NH_2 + CH_3\overset{O}{\overset{\|}{C}}O\overset{O}{\overset{\|}{C}}CH_3 \longrightarrow$

⬡$-NH\overset{O}{\overset{\|}{C}}CH_3 + CH_3CO_2H$ (eq 12-24)

d) ⬡$-NH_2 + CH_2\overset{\diagdown}{\underset{O}{-}}CH_2 \longrightarrow$ ⬡$-NHCH_2CH_2OH$

(eq 7-21)

e) ⬡$-NH_2 + $ ⬡$-SO_2Cl \xrightarrow{Na^+OH^-}$

⬡$-NH\overset{O}{\underset{O}{\overset{\|}{\underset{\|}{S}}}}$⬡$ + H_2O + Na^+Cl^- \xrightarrow[Na^+OH^-]{\text{excess}}$

⬡$-\overset{Na^+}{\underset{\underset{O}{\|}}{\overset{O}{\overset{\|}{\underset{|}{N}}}}}-S$⬡$ + H_2O$ (eq 12-30)

f) $\underset{\text{benzene ring}}{\bigcirc}-NH_2 + Na^+NO_2^- + 2\,H_2SO_4 \xrightarrow{0°}$

$\underset{\text{benzene ring}}{\bigcirc}-\overset{+}{N}\equiv N: \quad HSO_4^- + 2\,H_2O + Na^+HSO_4^- \quad \text{(eq 12-55)}$

g) $\underset{\text{benzene ring}}{\bigcirc}-NH_2 + \underset{\text{excess}}{CH_3I} \longrightarrow \left[\underset{\text{benzene ring}}{\bigcirc}-\overset{\overset{\displaystyle CH_3}{|}}{\underset{\underset{\displaystyle CH_3}{|}}{\overset{+}{N}}}-CH_3\right]\,I^- \quad \begin{array}{l}\text{(eq 12-8,}\\ \text{12-21)}\end{array}$

h) $\underset{\text{benzene ring}}{\bigcirc}-NH_2 + 3\,Br_2 \xrightarrow{H_2O} Br-\underset{\underset{\displaystyle Br}{|}}{\overset{\overset{\displaystyle Br}{|}}{\bigcirc}}-NH_2 + 3\,HBr$

(eq 12-39)

12-12. The reaction begins with nucleophilic attack by the amine on the carbonyl group of the anhydride (review sec 10.7, especially eq 10-28).

$$CH_3\overset{O}{\overset{||}{C}}O\overset{O}{\overset{||}{C}}CH_3 + H_2\ddot{N}CH_2CH_3 \longrightarrow CH_3\overset{O^-}{\underset{\underset{\displaystyle O}{\overset{\displaystyle |}{O\overset{||}{C}CH_3}}}{\overset{|}{C}}}-\overset{+}{N}H_2CH_2CH_3 \xrightarrow{\sim H^+}$$

$$CH_3-\overset{\overset{\displaystyle O-H}{|}}{\underset{\underset{\displaystyle O}{\overset{\displaystyle |}{\overset{\displaystyle O\overset{||}{C}CH_3}}}{C}}}-NHCH_2CH_3 \xrightarrow{-HOAc} CH_3\overset{O}{\overset{||}{C}}-\ddot{N}HCH_2CH_3$$

Even though the resulting amide has an unshared electron pair on nitrogen, it does not react with a second mole of acetic anhydride to become diacylated:

$$CH_3\overset{O}{\overset{||}{C}}O\overset{O}{\overset{||}{C}}CH_3 + CH_3\overset{O}{\overset{||}{C}}NHCH_2CH_3 \xrightarrow{\;\;\;\nrightarrow\;\;\;}$$

$$CH_3\overset{O}{\overset{||}{C}}-\underset{\underset{\displaystyle O=\overset{}{C}CH_3}{|}}{N}CH_2CH_3 + CH_3CO_2H$$

The reason is that amides are poor nucleophiles, due to the fact that the unshared electron pair on the nitrogen is delocalized (sec 10.8):

$$CH_3\overset{O}{\overset{\|}{C}}-\overset{..}{N}HCH_2CH_3 \longleftrightarrow CH_3\overset{O^-}{\overset{|}{C}}=\overset{+}{N}HCH_2CH_3$$

12-13. *a*) The formula for choline is given in sec 12.5b.

$$(CH_3)_3\overset{..}{N}: + CH_2-CH_2 \xrightarrow[\text{(eq 7-21)}]{\text{base}} (CH_3)_3\overset{+}{N}CH_2CH_2O^-$$

with the epoxide oxygen bridging the two CH_2 groups.

$$\big\updownarrow H_2O$$

$$[(CH_3)_3\overset{+}{N}CH_2CH_2OH]OH^-$$
choline

b) The desired compound is an insect repellant (sec 12.5c).

$$\underset{\text{m-toluic acid}}{\underset{\substack{| \\ CH_3}}{\overset{CO_2H}{\bigcirc}}} \xrightarrow[\text{(eq 10-31)}]{SOCl_2} \underset{\substack{| \\ CH_3}}{\overset{COCl}{\bigcirc}} \xrightarrow[\text{(eq 12-26)}]{(CH_3CH_2)_2NH}$$

$$\underset{\substack{| \\ CH_3}}{\overset{CON(CH_2CH_3)_2}{\bigcirc}} \quad + \text{ HCl}$$

12-14. The amines are 1°, 2°, and 3° respectively.

a) (sec 12.5d)

$$(CH_3)_2CHNH_2 + HONO \xrightarrow{0°}$$

$$(CH_3)_2CHOH + N_2\uparrow + H_2O \qquad \text{(eq 12-37)}$$
A gas will be evolved.

$$\underset{CH_3CH_2}{\overset{CH_3}{\diagdown}}NH + HONO \xrightarrow{0°}$$

$$\underset{CH_3CH_2}{\overset{CH_3}{\diagdown}}N-N=O + H_2O \quad \text{(eq 12-38)}$$
A yellow oily upper layer will appear.

$$(CH_3)_3N + HONO \xrightarrow{0°} \text{no observable reaction}$$

b) (sec 12.5c)

$(CH_3)_2CHNH_2$ + ⟨benzene⟩—SO_2Cl $\xrightarrow{\text{(eq 12-30)}}$

⟨benzene⟩—$SO_2NHCH(CH_3)_2$ + HCl

a precipitate, which is
soluble in base

$CH_3NHCH_2CH_3$ + ⟨benzene⟩—SO_2Cl $\xrightarrow{\text{(eq 12-31)}}$

⟨benzene⟩—$SO_2N(CH_3)CH_2CH_3$ + HCl

a precipitate, which is
insoluble in base

$(CH_3)_3N$ + ⟨benzene⟩—SO_2Cl $\xrightarrow{\text{(eq 12-32)}}$ no observable reaction

12-15. *a)* The alkyldiazonium ion is produced by elimination of water; several alternative paths are possible, only one of which is shown:

$$R-\overset{H}{\underset{H\ \ OH}{\overset{+}{N}}}-\overset{\cdot\cdot}{N}-\overset{\cdot\cdot}{O}:^{-} \underset{\sim H^{+}}{\rightleftharpoons} R-\overset{H}{\underset{\ OH}{N}}-\overset{\cdot\cdot}{N}-OH \xrightarrow{-H_2O} R-\overset{\cdot\cdot}{N}=\overset{\cdot\cdot}{N}-\overset{\cdot\cdot}{O}H$$

$$\Big\Updownarrow H^{+}$$

$$R-\overset{+}{N}\equiv N: \longleftrightarrow R-\overset{\cdot\cdot}{N}=\overset{+}{N}: \xleftarrow{-H_2O} R-\overset{\cdot\cdot}{N}=\overset{\cdot\cdot}{N}-\overset{+}{\underset{H}{O}}-H$$

an alkyl diazonium ion

b) $R-\overset{R'}{\underset{H}{N}}:\overset{\frown}{+}\overset{\cdot\cdot}{N}=\overset{\frown}{O}: \longrightarrow R-\overset{R'}{\underset{H\ \ OH}{\overset{+}{N}}}-\overset{\cdot\cdot}{N}-O^{-}$

$\Big\downarrow -H_2O$

$R-\overset{R'}{\underset{\cdot\cdot}{N}}-\overset{\cdot\cdot}{N}=O: \longleftrightarrow R-\overset{R'}{\underset{}{\overset{+}{N}}}=\overset{\cdot\cdot}{N}-\overset{\cdot\cdot}{O}\overline{:}$

a nitrosoamine

12-16.

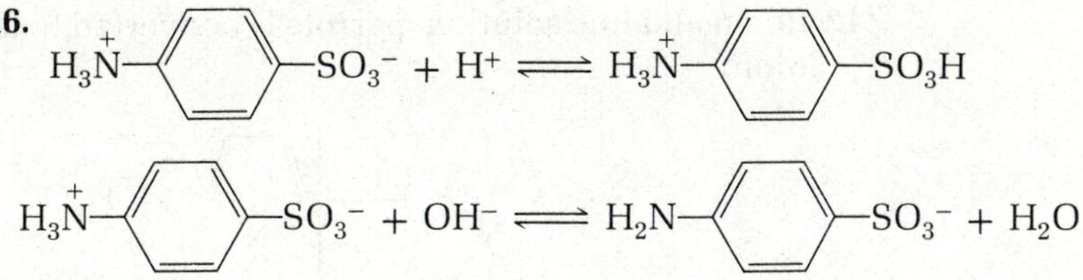

$$H_3\overset{+}{N}\!-\!\!\langle\ \rangle\!\!-\!SO_3^- + H^+ \rightleftharpoons H_3\overset{+}{N}\!-\!\!\langle\ \rangle\!\!-\!SO_3H$$

$$H_3\overset{+}{N}\!-\!\!\langle\ \rangle\!\!-\!SO_3^- + OH^- \rightleftharpoons H_2N\!-\!\!\langle\ \rangle\!\!-\!SO_3^- + H_2O$$

Substances which behave both as acids and as bases are called amphoteric.

12-17. *a)* $ClCH_2CH_2CH_2CH_2Cl \xrightarrow[\text{(eq 8-11)}]{Na^+CN^-} NC-CH_2CH_2CH_2CH_2-CN$

$$\text{(eq 10-10)} \downarrow H_2O, H^+ \qquad\qquad \downarrow \begin{array}{c} H_2/Pt \\ \text{(eq 12-46)} \end{array}$$

$$HO_2C-CH_2CH_2CH_2CH_2-CO_2H \leftarrow$$

$$H_2N-CH_2-CH_2CH_2CH_2CH_2-CH_2-NH_2$$

b)

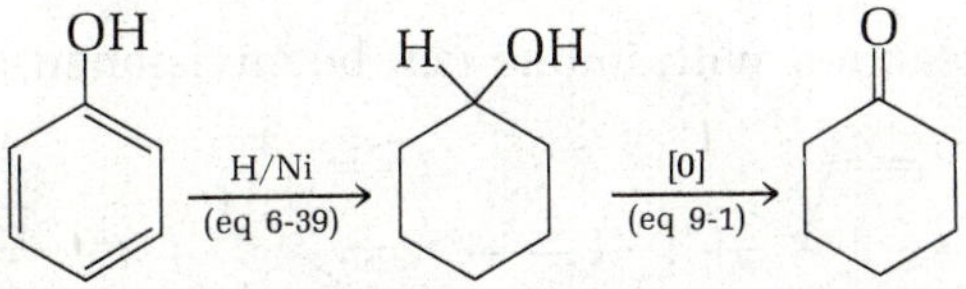

The cyclohexanone can now be used as in equations 12-45 and 12-46 to make adipic acid and hexamethylenediamine, the two nylon raw materials.

12-18. The necessary information is given in section 12.7.

a)

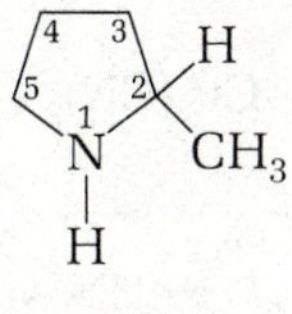

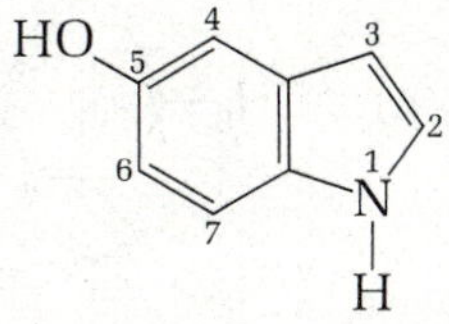

b)

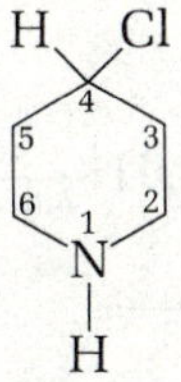

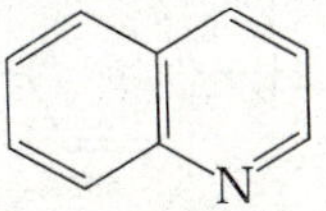

c)

d)

e)

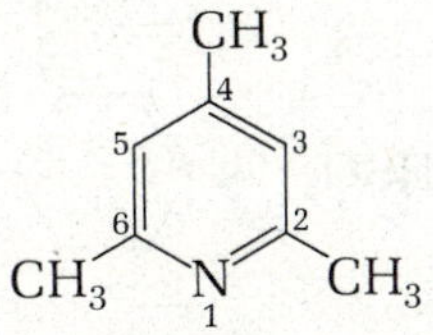

f)

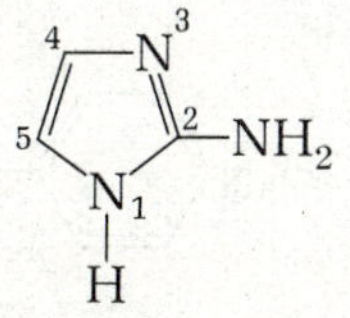

12-19. In alkaline solution, pyrrole is converted to the corresponding anion.

The anion is a good nucleophile and reacts readily with halogens (review sec 9.7b on the haloform reaction). It can react at any of the ring positions, since the negative charge is distributed over all of them through resonance.

The reaction with iodine can be envisioned as:

The iodoanion can react again with iodine, until the ring is completely substituted.

12-20. This question deals with material in section 12.8.

a) $2 \, CH_3-\!\!\!\bigcirc\!\!\!-\overset{+}{N}_2 \; HSO_4^- + Cu_2(CN)_2 \xrightarrow[\text{(eq 12-61)}]{}$

$2 \, CH_3-\!\!\!\bigcirc\!\!\!-CN + 2 \, N_2 + Cu_2(HSO_4)_2$

b) $CH_3-\!\!\!\bigcirc\!\!\!-\overset{+}{N}_2 \; HSO_4^- + H-\!OH \xrightarrow[\text{(eq 12-58)}]{\text{heat}}$

$CH_3-\!\!\!\bigcirc\!\!\!-OH + N_2 + H_2SO_4$

c) $CH_3-\!\!\!\bigcirc\!\!\!-\overset{+}{N}\!\!\equiv\!\!N\!: \; HSO_4^- + \bigcirc\!\!\!-N(CH_3)_2 \xrightarrow{OH^-}$

$CH_3-\!\!\!\bigcirc\!\!\!-N\!\!=\!\!N-\!\!\!\bigcirc\!\!\!-N(CH_3)_2$ (eq 12-64) *d)* CH$_3$—⟨ ⟩—$\overset{+}{\text{N}}_2$ HSO$_4^-$ + KI $\xrightarrow[\text{(sec 12.8b)}]{}$

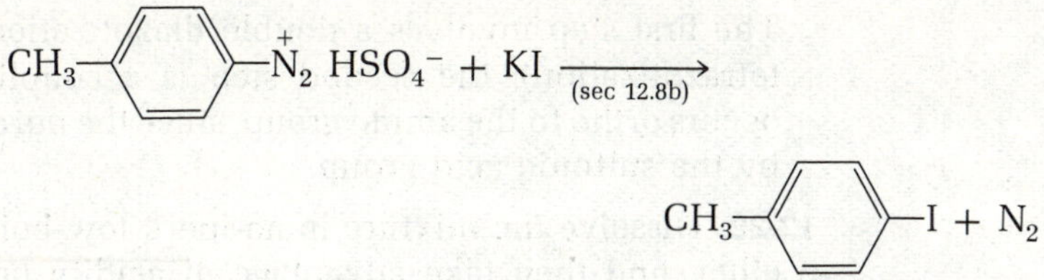

CH$_3$—⟨ ⟩—I + N$_2$

e) CH$_3$—⟨ ⟩—$\overset{+}{\text{N}}_2$ HSO$_4^-$ + Cu$_2$Cl$_2$ $\xrightarrow[\text{(eq 12-59)}]{}$

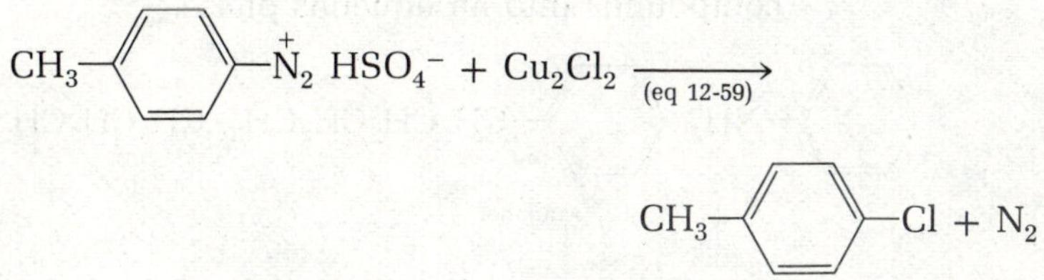

CH$_3$—⟨ ⟩—Cl + N$_2$

f) CH$_3$—⟨ ⟩—$\overset{+}{\text{N}}_2$ HSO$_4^-$ + HBF$_4$ $\xrightarrow[\text{(eq 12-60)}]{\text{heat}}$

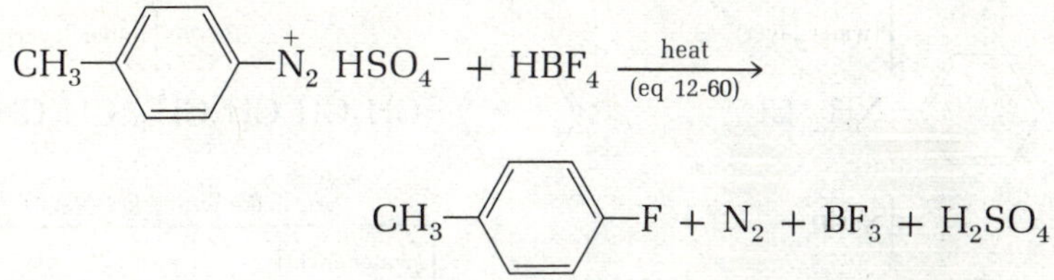

CH$_3$—⟨ ⟩—F + N$_2$ + BF$_3$ + H$_2$SO$_4$

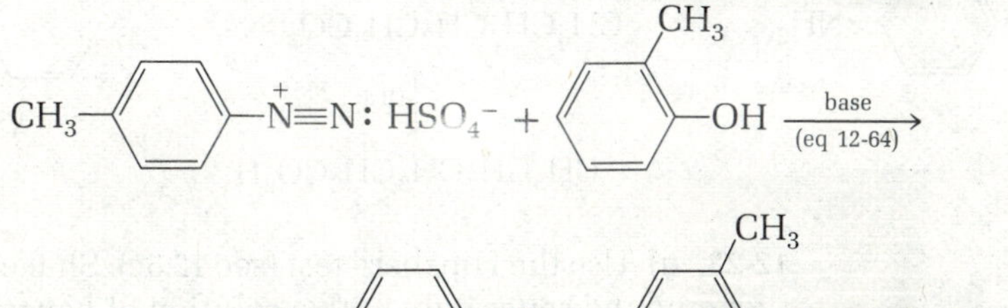

g) CH$_3$—⟨ ⟩—$\overset{+}{\text{N}}$≡N: HSO$_4^-$ + ⟨ ⟩(CH$_3$)—OH $\xrightarrow[\text{(eq 12-64)}]{\text{base}}$

CH$_3$—⟨ ⟩—N=N—⟨ ⟩(CH$_3$)—OH + H$_2$SO$_4$

h) CH$_3$—⟨ ⟩—$\overset{+}{\text{N}}_2$ HSO$_4^-$ $\xrightarrow[\text{D}_2\text{O}]{\text{H}_3\text{PO}_2}$ CH$_3$—⟨ ⟩—D + N$_2$

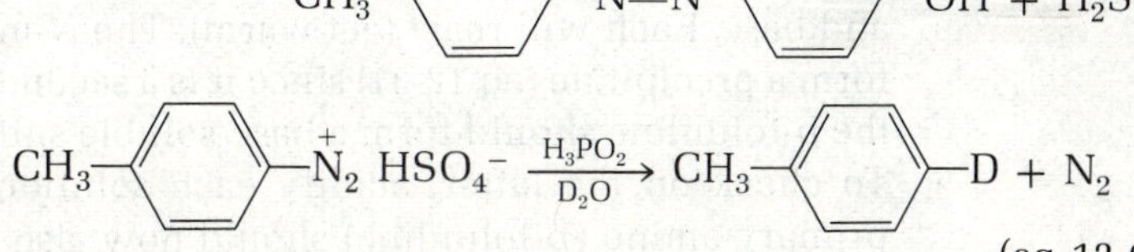

(eq 12-62)

12-21.

H$_2$N—⟨ ⟩—⟨ ⟩—NH$_2$ $\xrightarrow[\text{(eq 12-55)}]{\overset{\text{HONO}}{\text{H}^+}}$ $\overset{+}{\text{N}}_2$—⟨ ⟩—⟨ ⟩—$\overset{+}{\text{N}}_2$

benzidine

⟨NH$_2$⟩, base
SO$_3$H

(napthylamine sulfonic acid structure shown)

(azo product):

NH$_2$—(naphthalene)—N=N—⟨ ⟩—⟨ ⟩—N=N—(naphthalene with NH$_2$)

SO$_3$H   SO$_3$H

133

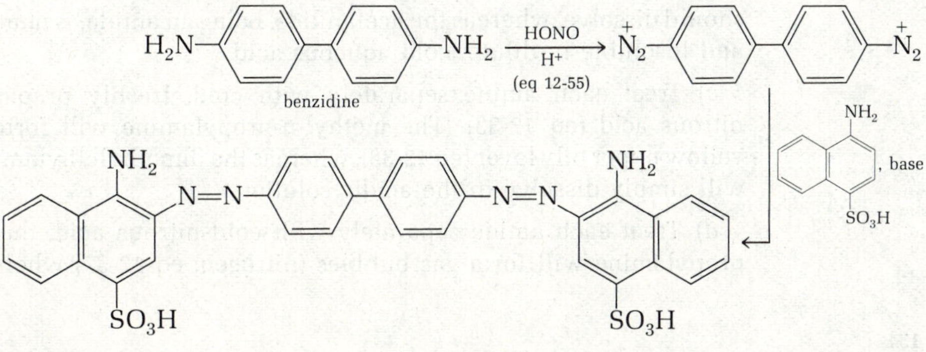

The first step involves a double diazotization (sometimes called tetrazotization); the second step is a coupling reaction, which occurs *ortho* to the amino group, since the *para* position is blocked by the sulfonic acid group.

12-22. Dissolve the mixture in an inert, low-boiling solvent such as ether, and then take advantage of acidity or basicity to extract compounds into an aqueous phase.

12-23. *a*) Use the Hinsberg test (sec 12.5c). Shake a few drops of each compound separately with a solution of benzenesulfonyl chloride and base. Each will react (get warm). The N-methylaniline should form a precipitate (eq 12-31) since it is a secondary amine, whereas the p-toluidine should form a base-soluble sulfonamide (eq 12-30). To check on the latter, acidify each solution; the one from the primary amine (*p*-toluidine) should now also form a precipate of a sulfonamide.

b) Add a little dilute HCl to each; the aniline, being a base, should dissolve, whereas the acetanilide, being an amide, is neutral and insoluble in dilute, cold aqueous acid.

c) Treat each amine separately with cold, freshly prepared nitrous acid (eq 12-33). The methyl-*n*-propylamine will form a yellow upper oily layer (eq 12-38), whereas the dimethylethylamine will simply dissolve in the acidic solution.

d) Treat each amine separately with cold nitrous acid; the *n*-propylamine will form gas bubbles (nitrogen; eq 12-37) whereas the methylethylamine will form an upper yellow oily layer (eq 12-38).

12-24. *a)*

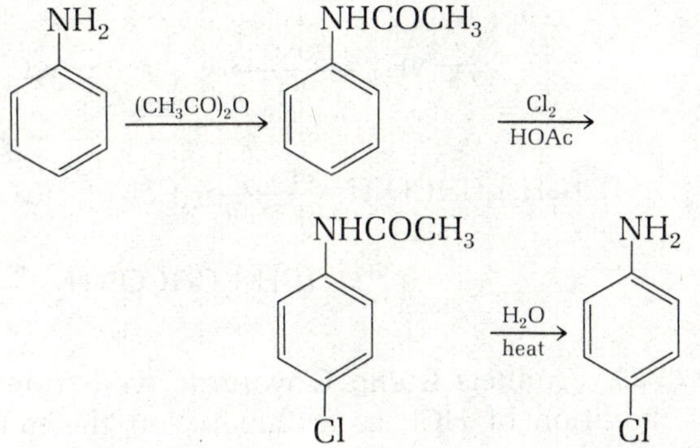

NH$_2$ $\xrightarrow{\text{(CH}_3\text{CO)}_2\text{O}}$ NHCOCH$_3$ $\xrightarrow[\text{HOAc}]{\text{Cl}_2}$

NHCOCH$_3$ (with Cl para) $\xrightarrow[\text{heat}]{\text{H}_2\text{O}}$ NH$_2$ (with Cl para) (eq 12-40)

b)

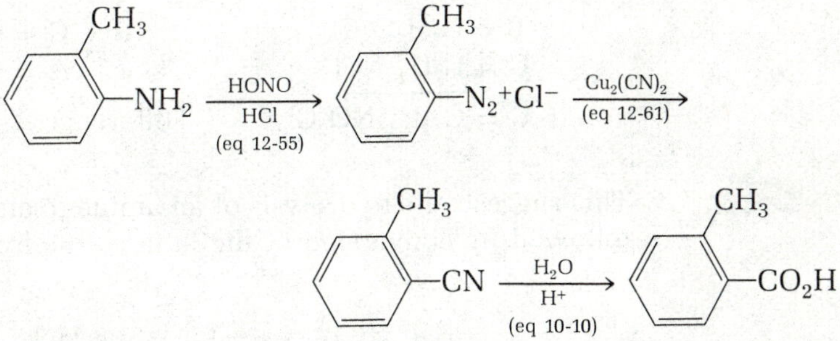

CH$_3$-ring-NH$_2$ $\xrightarrow[\text{HCl}]{\text{HONO}}$ (eq 12-55) CH$_3$-ring-N$_2^+$Cl$^-$ $\xrightarrow[\text{(eq 12-61)}]{\text{Cu}_2\text{(CN)}_2}$

CH$_3$-ring-CN $\xrightarrow[\text{H}^+]{\text{H}_2\text{O}}$ (eq 10-10) CH$_3$-ring-CO$_2$H

c)

CH$_3$-ring-NH$_2$ $\xrightarrow[\text{(eq 12-55)}]{\text{HONO, HCl}}$ CH$_3$-ring-N$_2^+$Cl$^-$ $\xrightarrow[\text{(eq 12-59)}]{\text{Cu}_2\text{Br}_2}$

CH$_3$-ring-Br $\xrightarrow[\text{(eq 10-3)}]{\text{KMnO}_4}$ CO$_2$H-ring-Br

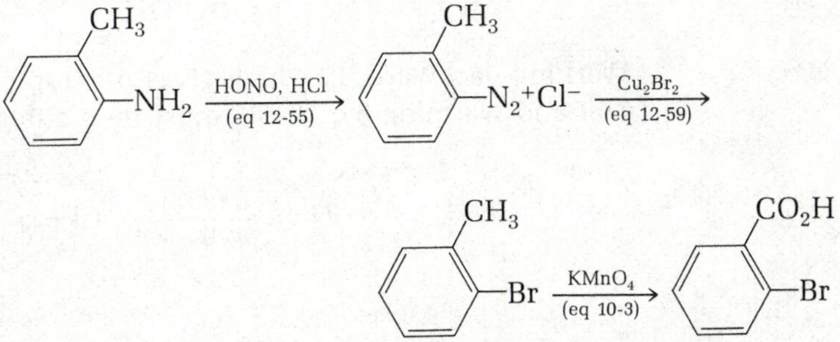

d) CH$_2$=CHCH$_2$CH$_3$ $\xrightarrow[\text{(eq 6-11, 6-12)}]{\text{H}_2\text{O} \atop \text{H}^+}$ CH$_3$CHCH$_2$CH$_3$ (with OH) $\xrightarrow[\text{(eq 9-1)}]{\text{K}_2\text{Cr}_2\text{O}_7}$

CH$_3$CCH$_2$CH$_3$ (with =O) $\xrightarrow[\text{(eq 9-29)}]{\text{NH}_2\text{OH}}$

CH$_3$CCH$_2$CH$_3$ (with =NOH) $\xrightarrow[\text{(eq 12-10)}]{\text{H}_2\text{/Pt}}$ CH$_3$CHCH$_2$CH$_3$ (with NH$_2$)

135

e)

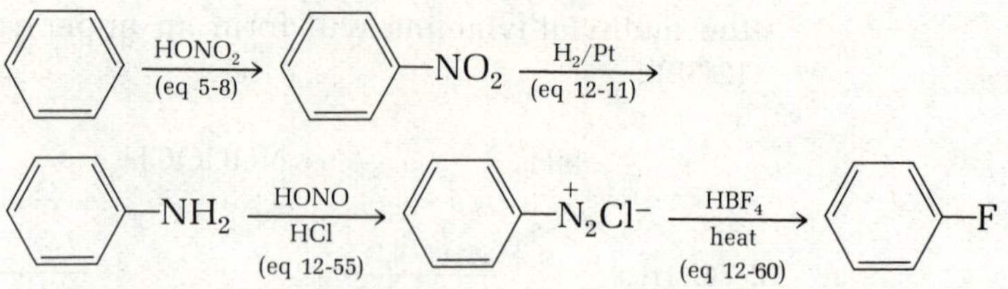

f) $(CH_3)_2CHCO_2H \xrightarrow[\text{(eq 10-31)}]{SOCl_2} (CH_3)_2CHCOCl \xrightarrow[\text{(eq 10-40)}]{NH_3}$

$$(CH_3)_2CHCONH_2 \xrightarrow[\text{(eq 12-12)}]{Br_2,\ NaOH} (CH_3)_2CHNH_2.$$

12-25. Products **B** and **C** were derived from **A** by hydrolysis and addition of HCl, as arithmetic on the molecular formulas will show:

$B = C_7H_5O_2Cl$	$B + C = C_{14}H_{15}NO_2Cl_2$
$C = C_7H_{10}NCl$	$A = C_{14}H_{12}NOCl$
$B + C = C_{14}H_{15}NO_2Cl_2$	Difference $=$ $\quad$ H_3 $\quad$ OCl
	$= H_2O + HCl$

This suggests the hydrolysis of an amide to an acid and an amine, followed by conversion of the latter to its hydrochloride:

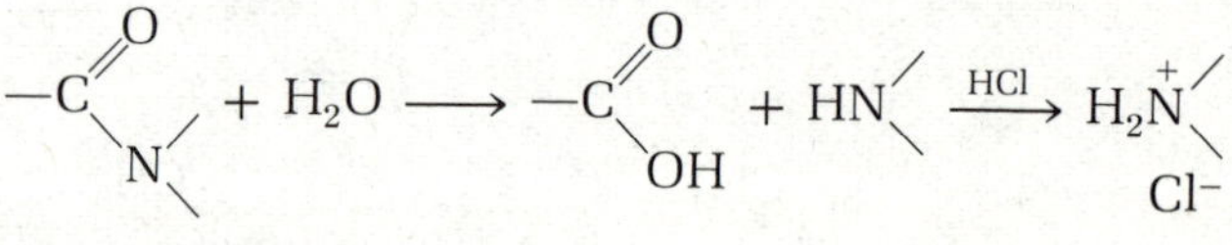

Working backward, **E**, which gives *p*-chlorophenol on diazotization and warming (eq 12-58), must be *p*-chloroaniline.

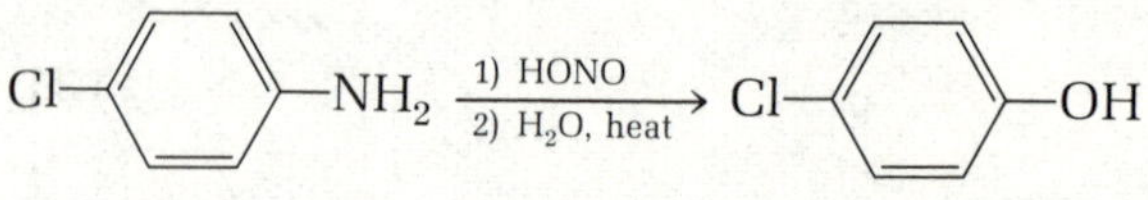

E
C_6H_6NCl

E was derived from **D** by the Hofmann rearrangement (sec 12.4c). **D** is therefore an amide (note that **D** differs from **E** by only one carbon and one oxygen).

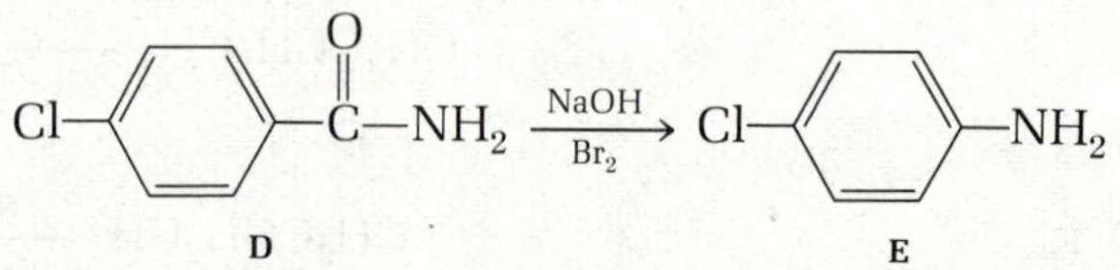

D $\qquad\qquad\qquad\qquad\qquad$ **E**

B must therefore be the corresponding acid.

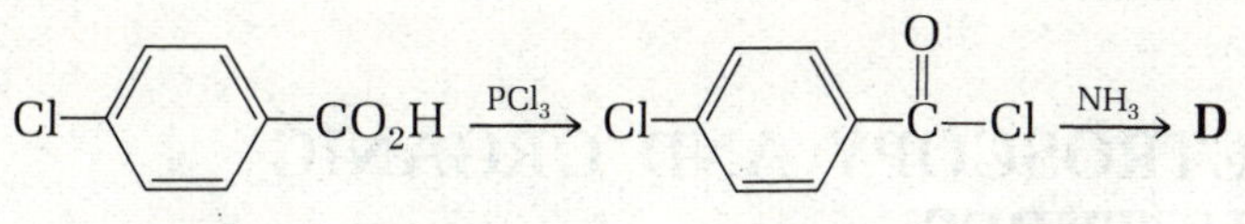

$$Cl--CO_2H \xrightarrow{PCl_3} Cl--\overset{\overset{\displaystyle O}{\|}}{C}-Cl \xrightarrow{NH_3} D$$

B
$C_7H_5O_2Cl$

Proceeding to the other hydrolysis product of **A,** the fact that **C** is water-soluble tends to confirm the idea that it is an ammonium salt (an amine hydrochloride). **F** must therefore be the free amine.

$$C_7H_{10}NCl + OH^- \longrightarrow C_7H_9N + H_2O + Cl^-$$
$$\phantom{C_7H_{10}NCl + OH^- \longrightarrow }\mathbf{F}$$

The remaining facts indicate that **F** is a secondary amine; the carbon-hydrogen ratio suggests that it is aromatic, and a satisfactory structure is N-methylaniline.

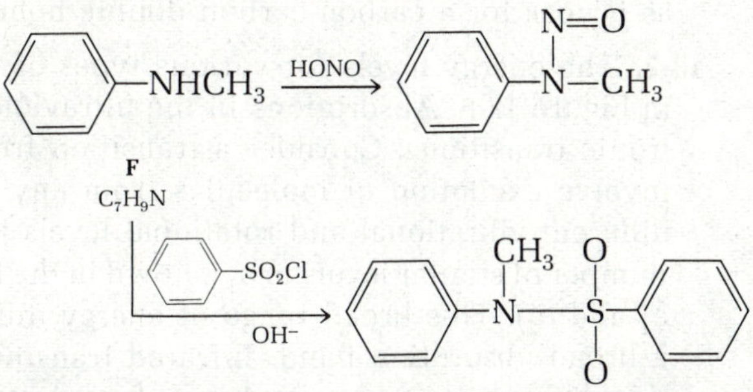

$$-NHCH_3 \xrightarrow{HONO} -\overset{\overset{\displaystyle N=O}{|}}{N}-CH_3$$

F
C_7H_9N

$$-SO_2Cl \xrightarrow{OH^-} -\overset{\overset{\displaystyle CH_3}{|}}{N}-\overset{\overset{\displaystyle O}{\|}}{\underset{\underset{\displaystyle O}{\|}}{S}}-$$

Putting the two structures (**F** and **B**) together, we get the following structure for **A,** and equation for its hydrolysis.

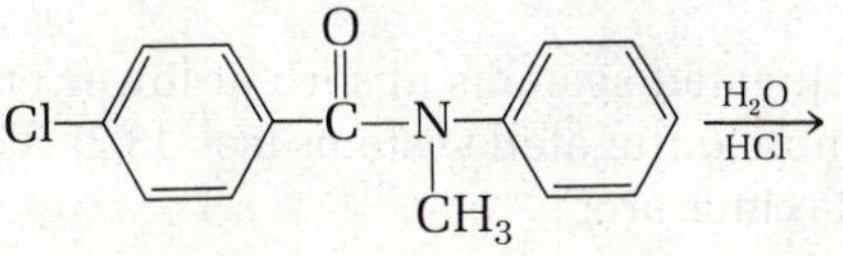

$$Cl--\overset{\overset{\displaystyle O}{\|}}{C}-\underset{\underset{\displaystyle CH_3}{|}}{N}- \xrightarrow[HCl]{H_2O}$$

A
$C_{14}H_{12}NOCl$

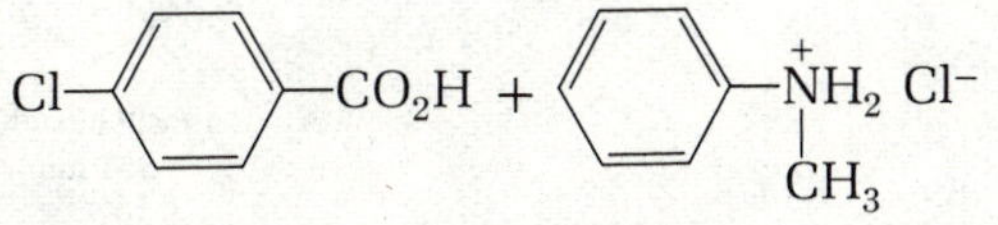

$$Cl--CO_2H + -\overset{\overset{\displaystyle +}{N}H_2}{\underset{\underset{\displaystyle CH_3}{|}}{}} Cl^-$$

$$\phantom{Cl--CO_2H + }\mathbf{B} \mathbf{C}$$

CHAPTER THIRTEEN

SPECTROSCOPY AND ORGANIC STRUCTURES

13-1. If we use the subscripts 3 and 2 to stand for triple and double bonds, respectively, we get

$$E_3 = h\nu_3; \; E_2 = h\nu_2$$

$$\frac{E_3}{E_2} = \frac{h\nu_3}{h\nu_2} = \frac{2180}{1650} = 1.32$$

The values 2180 and 1650 cm^{-1} are average values; it requires about one-third again more energy to stretch a carbon-carbon triple bond as it does for a carbon-carbon double bond.

13-2. The energy levels for various types of transitions are shown in Figure 13-5. Absorptions in the ultraviolet correspond to electronic transitions. Consider a transition from e_0 to e_1. This may involve excitation of molecules from any of a large number of different vibrational and rotational levels in e_0 to any of a large number of similar levels in e_1 (shown in the figure as lines between e_1 and e_2). This broad range of energy differences will result in a broad absorption band. Infrared transitions are mainly vibrational. Consider a transition in e_0 from v_1 to v_2. This may include molecules in various rotational levels in each of these states, but the range of energy differences is rather small, so the bands will be narrow.

13-3. In general, conjugated systems absorb at lower energy (longer wavelength) than nonconjugated systems (sec 13.2). The structures and absorption maxima are:

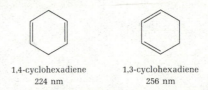

1,4-cyclohexadiene   1,3-cyclohexadiene
224 nm   256 nm

13-4. The only difference in the two structures is the position of the "middle" double bond; in each case this double bond is conjugated with the terminal double bond at the "right" end of the structure. As a first approximation, one would expect nearly identical spectra for the two structures, and in fact they are similar. However the alkyl substitution on the "middle" double bond in structure **B** is greater by one alkyl group than in structure **A**. Consequently one expects **B** to absorb at about 12–16 nm longer wavelength than **A** (compare with the chart of conjugated ketones in sec 13.2). Myrcene has structure **A**, whereas ocimene has structure **B**.

13-5. The two possible unsaturated ketones are

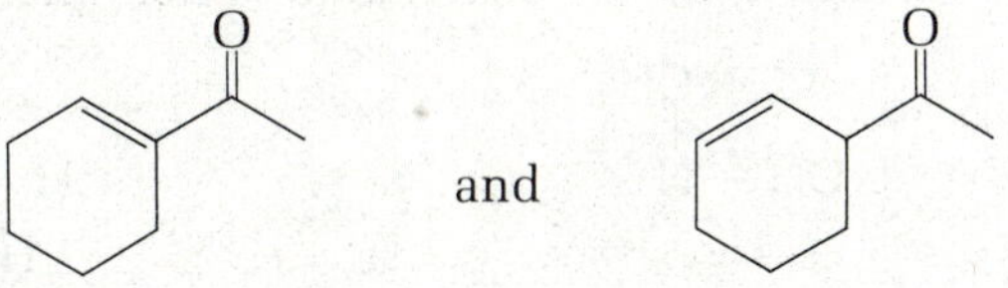

and

The predominant product, which absorbs at 236 nm, has the conjugated structure at the left. The carbon-carbon double bond is dialkyl-substituted, and the wavelength 236 nm corresponds very well to those of similar enones given in the chart in sec 13.2.

The minor product, being nonconjugated, should absorb at much shorter wavelength, perhaps near 200 nm.

13-6. In order for the phenyls and the central double bond to be maximally conjugated, the entire structure should be coplanar. This is possible for *trans*-stilbene, but in the *cis* isomer hydrogens in the *ortho* positions of the phenyl rings come very close to one another.

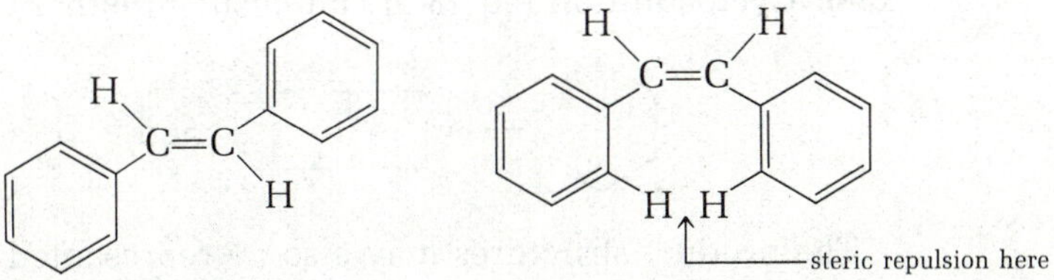

steric repulsion here

Consequently the *cis* isomer cannot be planar; this causes it to absorb at shorter wavelength (higher energy) and with less intensity than the *trans* isomer.

13-7. The main contributors to the resonance hybrid are:

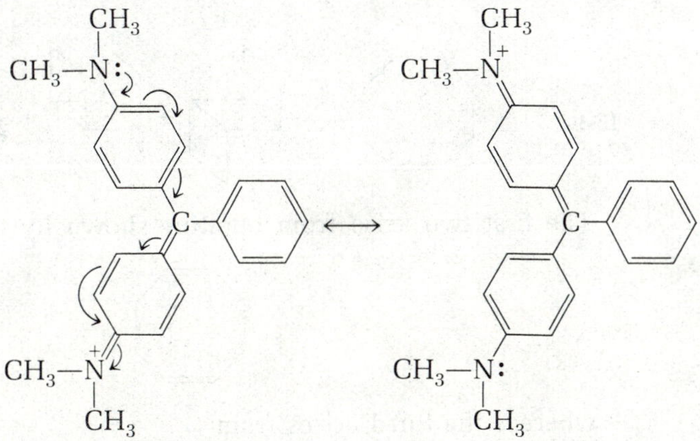

Structures with the positive charge at *ortho* and *para* carbons of the three phenyl rings are also possible.

In strongly acidic media ($pH < 0.5$) the neutral nitrogen can become protonated. This does not allow the above structures, though some delocalization is possible.

This effectively decreases the extent of conjugation and shortens the wavelength of absorption (green $\longrightarrow$ yellow).

13-8. Cycloaddition (eq 13-9) can occur in either of two ways:

The product structures may also be represented by the formulas

which convey their three-dimensionality.

13-9.

The first two arise from bonding shown by the dotted lines in

whereas the third arises from

13-10. Conjugation imparts some single bond character to the carbon-oxygen bond, through structures such as

$$C{=}C{-}C{=}\underset{..}{\overset{..}{O}}: \longleftrightarrow \overset{+}{C}{-}C{=}C{-}\underset{..}{\overset{..}{O}}:^{-}$$

This lowers the carbonyl absorption frequency, since the bond is somewhat easier to stretch.

13-11. As the solution of p-hydroxybenzaldehyde becomes more concentrated, *intermolecular* hydrogen bonding becomes important, involving structures such as

HO⟨benzene ring⟩—CH=O------HO⟨benzene ring⟩—CH=O

and

O=CH⟨benzene ring⟩—O—H------O⟨benzene ring⟩—CH=O
                                        |
                                        H

In the *ortho* isomer, however, there is strong *intramolecular* hydrogen bonding:

The geometry is such that this is a strong hydrogen bond, virtually unconverted to an intermolecular bond by increasing the concentration.

13-12. The mass of sulfur is twice that of oxygen. Consequently bonds to sulfur vibrate with lower frequency than corresponding bonds to oxygen. The C=S vibrational frequency comes in the range 1050–1200 cm^{-1}.

13-13. The peaks are assigned as follows:

$\delta 2.35$ — CH$_3$
$\delta 7.15$
$\delta 7.87$
$\delta 3.82$ — O—CH$_3$

In general, aromatic protons come at lower field if they are adjacent to electron-withdrawing substituents; this is the basis for distinguishing between the two sets of aromatic protons.

13-14. The ir band at 1725 cm^{-1} is due to a carbonyl group, probably a ketone. The quartet-triplet pattern in the nmr spectrum suggests an ethyl group. The compound is 3-pentanone:

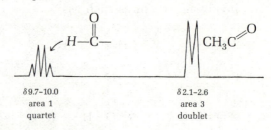

13-15. *a*) CH_3CHO

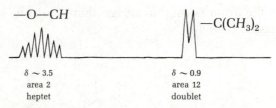

δ 9.7–10.0	δ 2.1–2.6
area 1	area 3
quartet	doublet

b) $(CH_3)_2CHOCH(CH_3)_2$

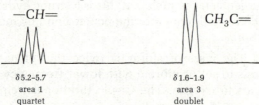

$\delta \sim 3.5$	$\delta \sim 0.9$
area 2	area 12
heptet	doublet

c) $CH_3CH{=}CCl_2$

—CH= $CH_3C{=}$

δ 5.2–5.7	δ 1.6–1.9
area 1	area 3
quartet	doublet

13-16. The first compound must have nine equivalent hydrogens. The only possible structure is *t*-butyl bromide:

$$\begin{array}{c} CH_3 \\ | \\ CH_3 \!-\! C \!-\! Br \\ | \\ CH_3 \end{array}$$

Its isomer has three different types of hydrogens, two of one kind, one unique, and six equivalent. The compound must be isobutyl bromide. The chemical shifts and spin-spin splitting pattern fit this structure:

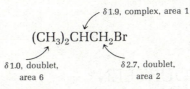

$\delta 1.9$, complex, area 1

$(CH_3)_2CHCH_2Br$

$\delta 1.0$, doublet, area 6

$\delta 2.7$, doublet, area 2

13-17. The quartet-triplet pattern suggests that the ten protons are present as two ethyl groups. This gives a partial structure $(CH_3CH_2)_2CO_3$. The chemical shift of the CH_2 groups ($\delta 4.15$) suggests that they are attached to the oxygen atoms. Finally, the carbonyl band at 1745 cm^{-1} suggests an ester function. The structure is diethyl carbonate:

$$CH_3CH_2O\overset{\overset{\textstyle O}{\|}}{C}OCH_2CH_3$$

13-18. Cleavage on either side of the carbonyl group would produce fragment ions with the indicated masses:

$$CH_3CH_2CH_2\overset{\overset{\textstyle O}{\|}}{C}CH_2CH_2CH_2CH_3$$
43 | 85

$$CH_3CH_2CH_2\overset{\overset{\textstyle O}{\|}}{C}CH_2CH_2CH_2CH_3$$
71 | 57

13-19. The absence of a band at 3500 cm^{-1} indicates that there are no hydroxyl groups; the absence of a band at 1720 cm^{-1} indicates that the compound is not an aldehyde or ketone. This suggests that the oxygen function is probably an ether. Possible structures are

$$CH_3OCH=CH_2$$

$$\begin{array}{c} O \\ CH_2 \quad CH_2 \\ CH_2 \end{array}$$

$$\begin{array}{c} O \\ CH_2-CH-CH_3 \end{array}$$

An nmr spectrum would readily distinguish amongst these possibilities.

13-20. The nmr peak at $\delta 7.4$ with an area of 5 suggests that the compound may have a phenyl group, C_6H_5-. If so, this would account for 77 of the 102 mass units. This leaves only 25 mass units, one of which must be a hydrogen (for the nmr peak at $\delta 3.08$). The other 24 units must be two carbon atoms since the compound is a hydrocarbon (no other elements present except C and H).

Phenylacetylene fits all the data:

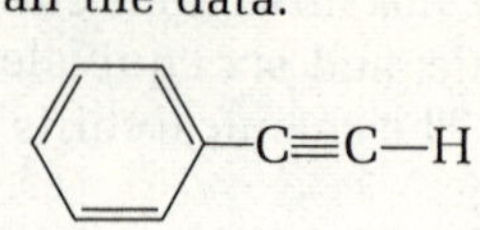

CHAPTER FOURTEEN

BIFUNCTIONAL COMPOUNDS

14-1. a) $CH_3CH_2CHCO_2H$
  |
  Br

b) CH_3C-CCH_3
     ‖   ‖
     O O
(also called biacetyl)

c)

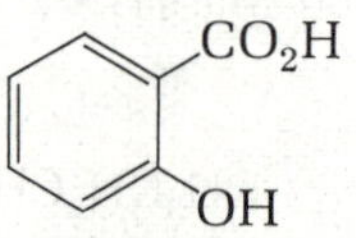

d)
   O
   ‖
$CH_3CCH_2CO_2CH_2CH_3$

e) $CO_2CH_2CH_3$
              /
CH_3CH_2CH
              \
                 $CO_2CH_2CH_3$

The "diethyl" part of the name indicates that the compound is a di-ester. The corresponding acid is ethylmalonic acid, a dicarboxylic acid.

f) CH_2CO_2H
   |
   CH_2CO_2H

g) $CH_3CHCO_2CH_3$
   |
   OH

h)

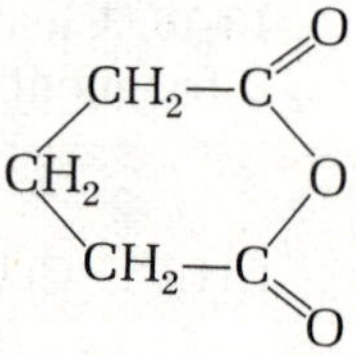

i)

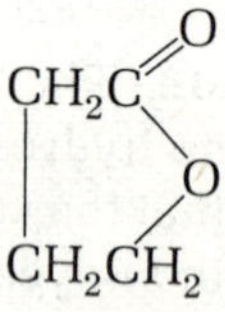

j) $HO-CHCO_2CH_2CH_3$
   |
   $CH_2CO_2CH_2CH_3$

This is the diethyl ester of malic acid.

14-2. a) γ-hydroxybutyric acid

b) methylmalonic acid

c) α-methylsuccinic acid

d) β-ketobutyric acid

e) glyoxal

f) pyruvic acid
   (or α-ketopropionic acid)

g) diethyl malonate
(or malonic ester)

h) 2,4-pentanedione

i) 4-bromosalicylic acid
(number from the carboxyl group)

j) methyl glycolate

14-3. *a)*

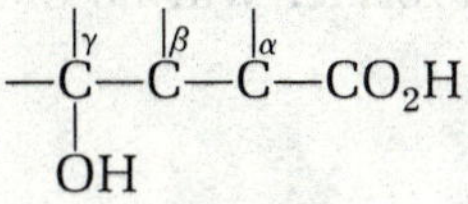

$$-\overset{|\gamma}{\underset{\overset{|}{OH}}{C}}-\overset{|\beta}{C}-\overset{|\alpha}{C}-CO_2H \qquad \text{(sec 14.2a)}$$

b)

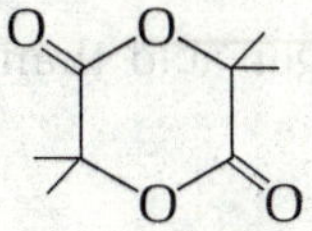

(eq 14-27)

c)

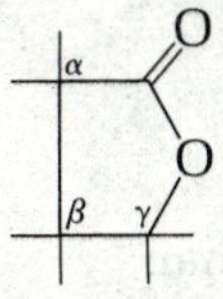

(eq 14-25)

d)

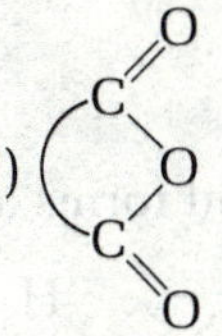

(sec 14.1a)

e)

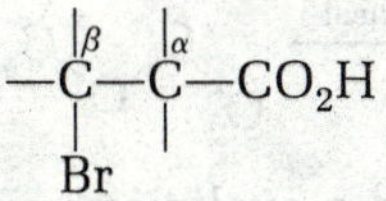

$$-\overset{|\beta}{\underset{\overset{|}{Br}}{C}}-\overset{|\alpha}{C}-CO_2H$$

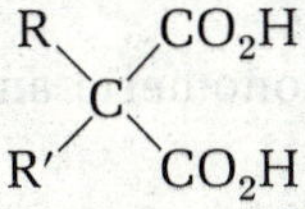

f)

$$\overset{R}{\underset{R'}{}}C\overset{CO_2H}{\underset{CO_2H}{}} \qquad \text{(sec 14.1b)}$$

g)

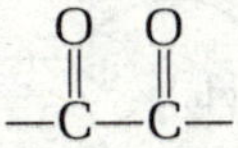

$$-\overset{O}{\overset{\|}{C}}-\overset{O}{\overset{\|}{C}}- \qquad \text{(sec 14.1c)}$$

h)

$$-\overset{|\beta}{\underset{\overset{|}{OH}}{C}}-\overset{|\alpha}{C}-\overset{O}{\overset{\|}{C}}-H \qquad \text{(sec 9.7a)}$$

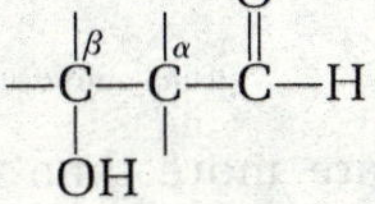

14-4. Consider K_{a_1} first. The carboxyl group is an electron-withdrawing substituent, and therefore will enhance the acidity of a nearby carboxyl group (sec 10.2). Since the substituent carboxyl is closer in phthalic acid than in terephthalic acid to the carboxyl group which ionizes, the former is the stronger acid.

Consider now K_{a_2}. The substituent is a carboxylate ion, $-CO_2^-$. Since the substituent is negatively charged, it will resist the formation of another nearby negative charge. Since the $-CO_2^-$ group is closer in phthalic acid than in terephthalic acid to the $-CO_2H$ group which ionizes, the former will be the weaker acid.

To sum up:

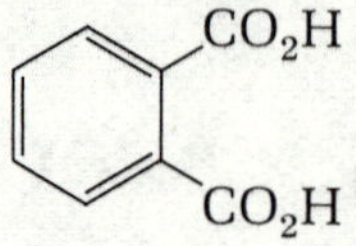

is a stronger acid than

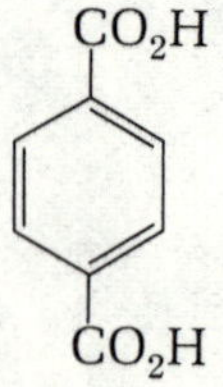

but

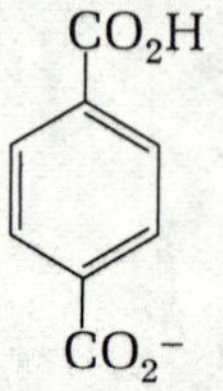

is a stronger acid than

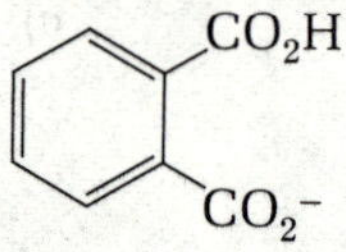

14-5. *cis*-1,2-Cyclopropanecarboxylic acid forms a cyclic anhydride:

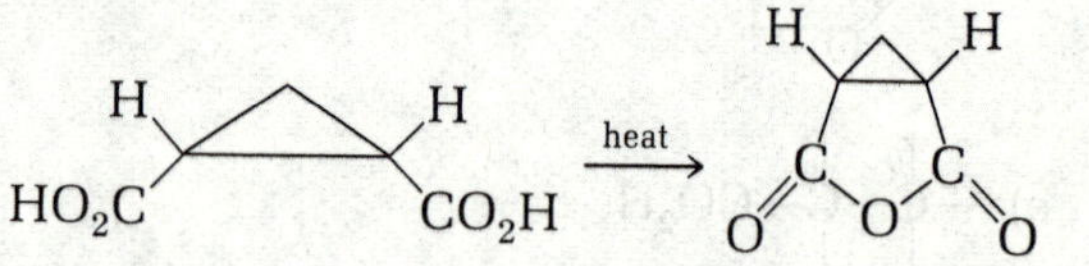

(cf. eq 14-3)

The *trans* isomer cannot form a cyclic anhydride (cf. eq 14-6); it may form a polymeric anhydride.

When the ring is larger, the monomeric anhydride is possible for both isomers:

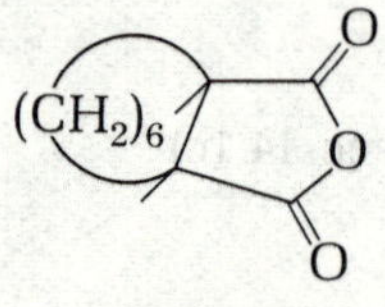

cis-anhydride 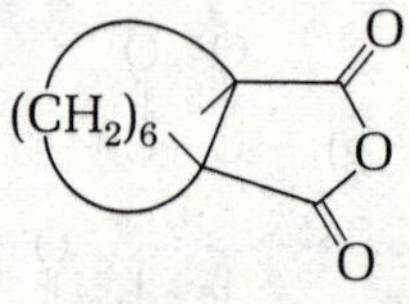   trans-anhydride

Six methylene (CH_2) groups are more than sufficient for being fused either *cis* or *trans* relative to the five-membered anhydride ring.

14-6. *a*) The *cis* isomer can form a cyclic anhydride, best shown using conformational formulas:

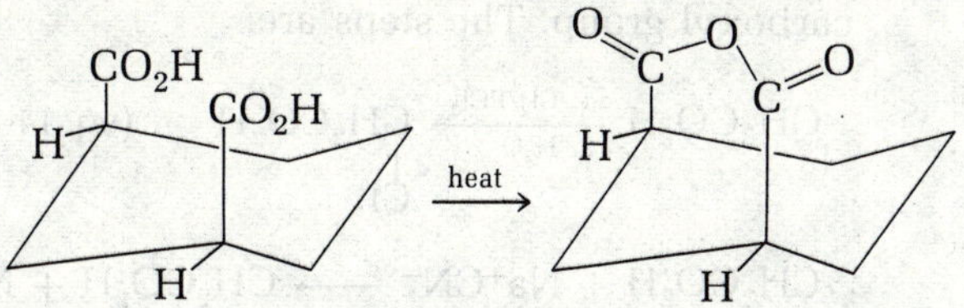

(eq 14-4)

b) 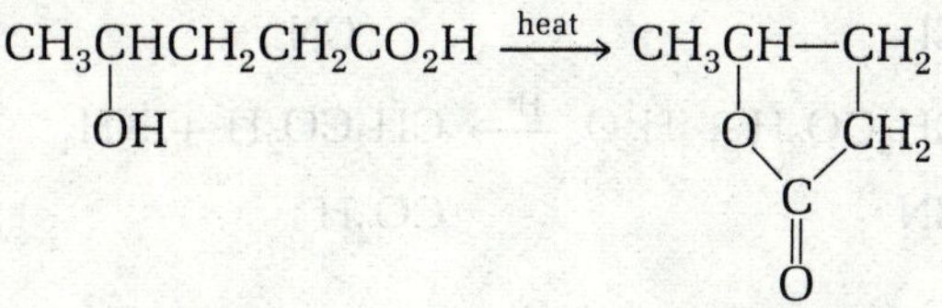 $CH_3CHCH_2CH_2CO_2H$ $\xrightarrow{\text{heat}}$ (eq 14-25)

γ-methyl-γ-butyrolactone

c) 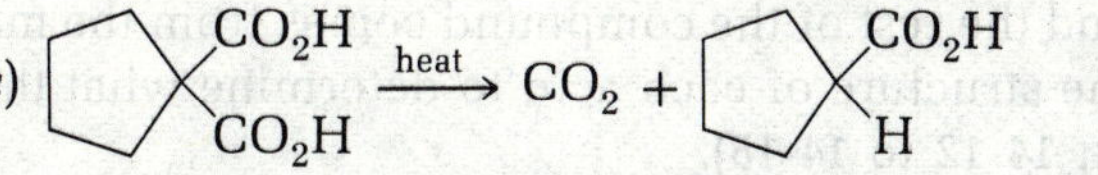 $\xrightarrow{\text{heat}}$ CO_2 + (eq 14-8)

The reaction is entirely analogous to the decarboxylation of a malonic acid.

d) 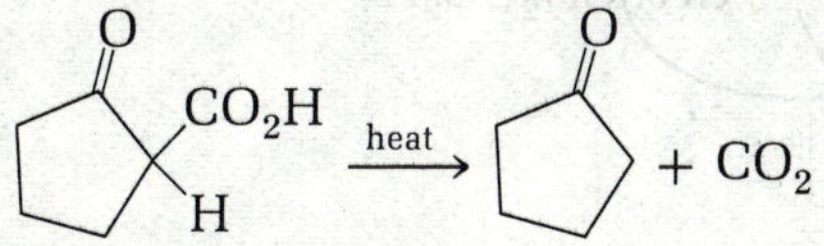 $\xrightarrow{\text{heat}}$ + CO_2 (eq 14-32)

The reaction is analogous to the decarboxylation of acetoacetic acid; it is general for β-keto acids.

e) 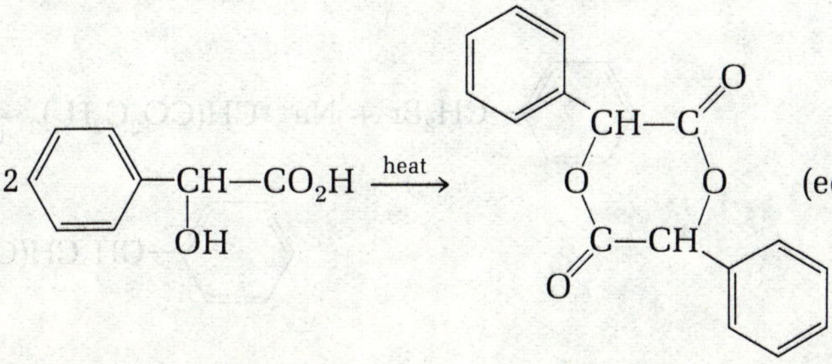 2 $\xrightarrow{\text{heat}}$ (eq 14-27)

The product is a lactide.

f) 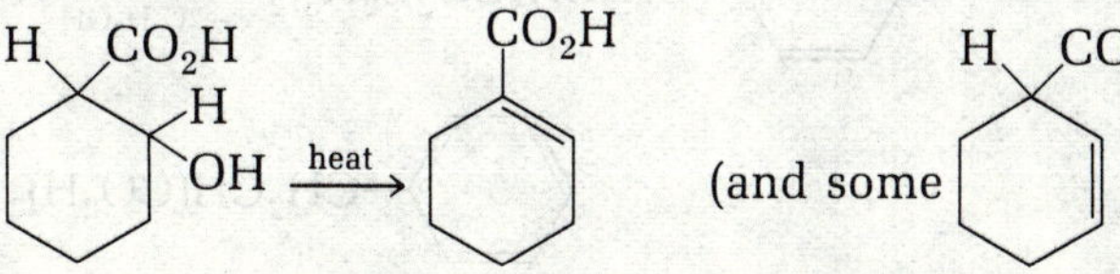 $\xrightarrow{\text{heat}}$ (and some) (eq 14-28)

14-7. The problem $CH_3CO_2H \longrightarrow CH_2(CO_2H)_2$ requires the replacement of one of the methyl hydrogens in acetic acid by a carboxyl group. The steps are:

$$CH_3CO_2H \xrightarrow[\text{2. H}_2\text{O}]{\text{1. Cl}_2/\text{PCl}_3} \underset{\underset{Cl}{|}}{CH_2CO_2H} \qquad \text{(eq 14-22)}$$

$$\underset{\underset{Cl}{|}}{CH_2CO_2H} + Na^+CN^- \longrightarrow \underset{\underset{CN}{|}}{CH_2CO_2H} + Na^+Cl^- \qquad \text{(eq 8-11)}$$

$$\underset{\underset{CN}{|}}{CH_2CO_2H} + H_2O \xrightarrow{H^+} \underset{\underset{CO_2H}{|}}{CH_2CO_2H} + NH_4^+ \qquad \text{(eq 10-10)}$$

14-8. The malonic ester synthesis can be used to prepare acids of the type $R_1R_2CHCO_2H$. The R groups come from the alkyl halide, and the rest of the compound comes from the malonic ester. Write the structure of each acid to determine what the R groups are (in eq 14-12 to 14-15).

a)

$$CH_2(CO_2C_2H_5)_2 + Na^+{}^-OC_2H_5 \xrightarrow[\text{(eq 14-11)}]{}$$

$$Na^+{}^-CH(CO_2C_2H_5)_2 + C_2H_5OH$$

$$-CH_2Br + Na^+{}^-CH(CO_2C_2H_5)_2 \xrightarrow[\text{(eq 14-12)}]{}$$

$$-CH_2CH(CO_2C_2H_5)_2 + Na^+Br^-$$

$$-CH_2CH(CO_2C_2H_5)_2 \xrightarrow[-2C_2H_5OH]{\text{NaOH}}$$

$$-CH_2CH(CO_2H)_2 \xrightarrow[-CO_2]{\text{heat}}$$

$$-CH_2CH_2CO_2H \qquad \text{(eq 14-13)}$$

b) $CH_3CH_2CHCO_2H$
     $\underset{\uparrow}{CH_3}$

two R groups

The order in which the two R groups are introduced is not important in this case; if the R groups are bulky, the larger group is usually introduced first.

The first equation is the same as in part *a*. Then

$$CH_3CH_2Br + Na^{+-}CH(CO_2C_2H_5)_2 \xrightarrow[(eq\ 14\text{-}12)]{}$$

$$CH_3CH_2CH(CO_2C_2H_5)_2 + Na^+Br^-$$

$$CH_3CH_2CH(CO_2C_2H_5)_2 + Na^{+-}OC_2H_5 \xrightarrow[(eq\ 14\text{-}15)]{}$$

$$CH_3CH_2\overset{-}{C}(CO_2C_2H_5)_2 + C_2H_5OH$$
$$Na^+$$

$$CH_3Br + Na^{+-}\underset{CH_2CH_3}{\overset{|}{C}}(CO_2C_2H_5)_2 \xrightarrow[(eq\ 14\text{-}15)]{} CH_3\underset{CH_2CH_3}{\overset{|}{C}}(CO_2C_2H_5)_2$$

$$\Big\downarrow \begin{array}{l} Na^+OH^- \\ (eq\ 14\text{-}13) \end{array}$$

$$CH_3\underset{CH_2CH_3}{\overset{|}{C}}HCO_2H \xleftarrow[(eq\ 14\text{-}13)]{\underset{-CO_2}{heat}} CH_3\underset{CH_2CH_3}{\overset{|}{C}}(CO_2H)_2$$

c) The sequence is the same as in part *b*, except that the alkyl halide in each case is cyclopentyl bromide (or chloride). The product is

d) The sequence is the same as in part *a*, except that the alkyl halide is allyl bromide, $CH_2{=}CHCH_2Br$ (or chloride). The product is

$$R{\equiv}(CH_2{=}CHCH_2)CH_2CO_2H$$

14-9. If excess malonic ester is used, the product will be adipic acid; if not, cyclopropanecarboxylic acid. The equations are

$$CH_2(CO_2C_2H_5)_2 + Na^{+-}OC_2H_5 \xrightarrow[(eq\ 14\text{-}11)]{}$$

$$Na^{+-}CH(CO_2C_2H_5)_2 + C_2H_5OH$$

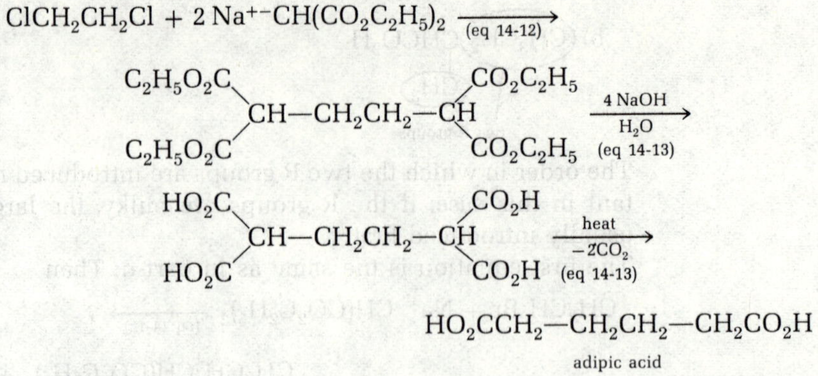

$$ClCH_2CH_2Cl + 2\,Na^{+-}CH(CO_2C_2H_5)_2 \xrightarrow[\text{(eq 14-12)}]{}$$

$$\underset{C_2H_5O_2C}{\overset{C_2H_5O_2C}{>}}CH-CH_2CH_2-CH\underset{CO_2C_2H_5}{\overset{CO_2C_2H_5}{<}} \xrightarrow[\text{H}_2\text{O}\;\;\text{(eq 14-13)}]{4\,\text{NaOH}}$$

$$\underset{HO_2C}{\overset{HO_2C}{>}}CH-CH_2CH_2-CH\underset{CO_2H}{\overset{CO_2H}{<}} \xrightarrow[-2CO_2\;\;\text{(eq 14-13)}]{\text{heat}}$$

$$HO_2CCH_2-CH_2CH_2-CH_2CO_2H$$

adipic acid

If less malonic ester is used, the following reactions can occur:

$$ClCH_2CH_2Cl + Na^{+-}CH(CO_2C_5H_5)_2 \xrightarrow[\text{(eq 14-12)}]{}$$

$$ClCH_2CH_2CH(CO_2C_2H_5)_2$$

In the presence of more base:

$$ClCH_2CH_2CH(CO_2C_2H_5)_2 + Na^{+-}OC_2H_5 \xrightarrow[\text{(eq 14-11)}]{}$$

$$ClCH_2CH_2\underset{Na^+}{\overset{}{\bar{C}}}(CO_2C_2H_5)_2 + C_2H_5OH$$

$$\underset{Na^+}{Cl-CH_2CH_2\bar{C}(CO_2C_2H_5)} \xrightarrow[\text{(eq 14-12)}]{\overset{\text{intramolecular}}{S_N2\text{ displacement}}} \underset{CH_2}{\overset{CH_2}{>}}C(CO_2C_2H_5)_2$$

$$\downarrow \begin{array}{l}\text{NaOH}\\ \text{(eq 14-13)}\end{array}$$

$$\underset{CH_2}{\overset{CH_2}{>}}\underset{CO_2H}{\overset{H}{C}} \xleftarrow[-CO_2\atop\text{(eq 14-13)}]{\text{heat}} \underset{CH_2}{\overset{CH_2}{>}}C(CO_2H)_2 \longleftarrow$$

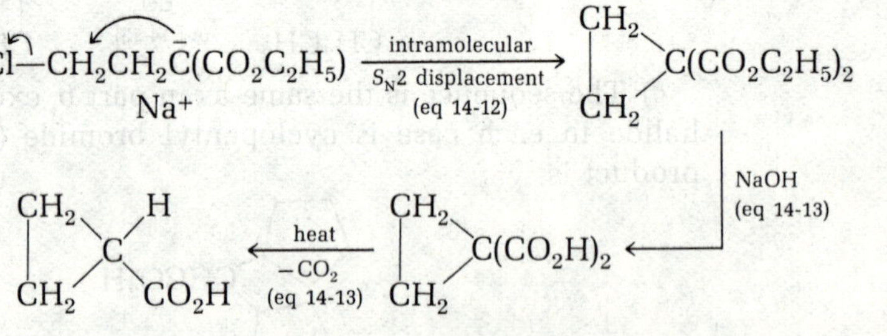

14-10. Consider the resonance contributors to the corresponding anion in each case. With acetylacetone the ion is symmetric, and each oxygen carries an equal amount of negative charge:

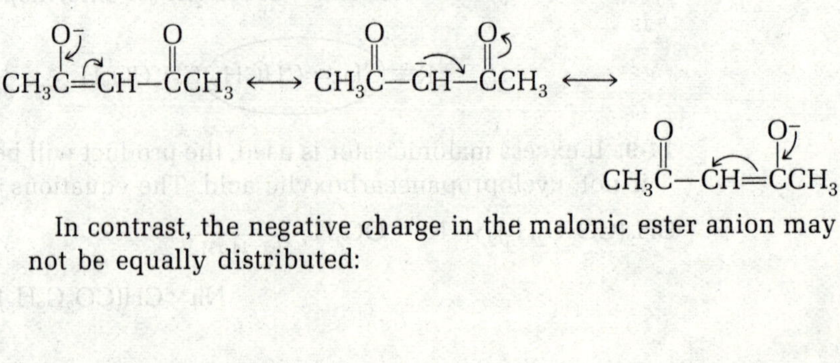

$$\underset{}{\overset{O^-\quad\;\;O}{CH_3C=CH-CCH_3}} \longleftrightarrow \underset{}{\overset{O\quad\quad O}{CH_3C-\bar{C}H-CCH_3}} \longleftrightarrow$$

$$\underset{}{\overset{O\quad\quad O^-}{CH_3C-CH=CCH_3}}$$

In contrast, the negative charge in the malonic ester anion may not be equally distributed:

$$CH_3\overset{\overset{\displaystyle O^-}{|}}{C}=CH-\overset{\overset{\displaystyle O}{||}}{C}-\ddot{O}C_2H_5 \longleftrightarrow CH_3\overset{\overset{\displaystyle O}{||}}{C}-\overset{..}{C}H-\overset{\overset{\displaystyle O}{||}}{C}-\ddot{O}C_2H_5 \longleftrightarrow$$

$$CH_3\overset{\overset{\displaystyle O}{||}}{C}-CH=\overset{\overset{\displaystyle O^-}{|}}{C}-\ddot{O}C_2H_5$$

The contributor at the right may be less important, because of the unshared electron pairs on the ethoxy group.

Put another way, the carboethoxy group is not as electron-withdrawing as the acetyl group, due to contributions from structures such as

$$CH_3\overset{\overset{\displaystyle O}{||}}{C}CH_2\overset{\overset{\displaystyle O}{||}}{C}-\ddot{O}C_2H_5 \longleftrightarrow CH_3\overset{\overset{\displaystyle O}{||}}{C}CH_2\overset{\overset{\displaystyle O^-}{|}}{C}=\overset{+}{O}C_2H_5$$

See also the answer to Problem 10-13a.

14-11. The β-diketone is in equilibrium with its enol (eq 14-19); both species give rise to peaks in the nmr spectrum.

$$CH_3\overset{\overset{\displaystyle O}{||}}{C}CH_2\overset{\overset{\displaystyle O}{||}}{C}CH_3 \rightleftharpoons$$

δ3.6   δ2.0

Equilibration between the two enols, as shown, makes the methyls in the enol equivalent.

14-12. In small rings, the carbonyls are coplanar and cisoid:

small rings

In large rings, the carbonyls are coplanar but transoid:

large rings

151

In rings of intermediate size (7 and 8 carbons) the two carbonyl groups are twisted and not coplanar. Since the absorption is at longer wavelength the more conjugated (planar), the two π systems (sec 13.2), the small ring and large ring α-diketones absorb at longest wavelength. Since large rings are more "floppy" than small rings, the shift is larger (466 nm) in the small rings than in the large rings (384 nm).

14-13. The precise mechanism is not yet known; the one which is presented below is plausible, but alternative mechanisms in which protons may be added or lost at other stages are also possible.

14-14. *a)* See eq 15-3.

b) See eq 14-24.

c) $CH_3\overset{\displaystyle O}{\underset{\displaystyle \|}{C}}CO_2H \xrightarrow{\text{LiAlH}_4} CH_3\underset{\displaystyle OH}{\underset{\displaystyle |}{C}}HCO_2H$

Some of the lithium aluminum hydride is wasted in forming the salt of the acid, but the acid is recovered at the end of the reaction, when water is added. Other reducing agents such as H_2/Ni can also be used (sec 9.6h).

d) $CH_3CHCH_2OH \xrightarrow[H^+]{K_2Cr_2O_7} CH_3CCO_2H$
 $\quad\quad |$ $\quad\quad\quad\quad\quad\quad\quad\quad\quad ||$
 $\quad\;\, OH \quad\quad\quad\quad\quad\quad\quad\quad\quad O$

(see sec 9.4a and 10.4a)

Both hydroxyls are oxidized.

14-15. The method is analogous to that described by equation 14-24 for lactic acid.

$$\text{Ph--CHO} \xrightarrow{\text{HCN}} \underset{\substack{\text{benzaldehyde}\\\text{cyanohydrin}}}{\text{Ph--}\overset{OH}{\underset{|}{CH}}\text{--CN}} \xrightarrow[H^+]{H_2O}$$

$$\text{Ph--}\overset{OH}{\underset{|}{CH}}\text{--CO}_2H$$

14-16.

$$\begin{array}{c} CH_3CH_2 \quad\quad CO_2C_2H_5 \\ \diagdown \diagup \\ C \\ \diagup \diagdown \\ CH_3CH_2 \quad\quad CO_2C_2H_5 \end{array} + \begin{array}{c} H_2N \\ \diagdown \\ C{=}O \\ \diagup \\ H_2N \end{array} \longrightarrow$$

$$\begin{array}{c} O \\ || \\ CH_3CH_2 \quad\quad C{-}NH \\ \diagdown \diagdown \\ C C{=}O + 2C_2H_5OH \\ \diagup \diagup \\ CH_3CH_2 \quad\quad C{-}NH \\ || \\ O \end{array}$$

The reaction involves nucleophilic attack by the urea nitrogens on the carbonyl groups of the diester. Compare this with equation 10–24.

14-17. Equations 14-25, 14-27, and 14-29 involve nucleophilic attack by a hydroxyl group on a carbonyl carbon atom (eq 10-28). Equation 14-28 is an acid-catalyzed elimination (sec 3.5a).

a)

$$\begin{array}{c} \overset{\displaystyle O}{} \\ \diagup\diagup \\ CH_2{-}C{-}OH \\ | \quad\quad :\ddot{O}H \\ CH_2{-}\ddot{C}H_2 \end{array} \longrightarrow \begin{array}{c} O^- \\ | \\ CH_2{-}C{-}OH \\ | \quad\quad \overset{+}{:}O{-}H \\ CH_2{-}CH_2 \end{array} \underset{-H^+,\,+H^+}{\rightleftharpoons} \begin{array}{c} O{-}H \\ \diagdown \\ CH_2{-}C{-}OH \\ | \quad\quad \diagdown \\ CH_2{-}CH_2 \quad O \end{array}$$

$$\begin{array}{c} O \\ || \\ CH_2{-}C \\ | \quad\quad\quad \diagdown \\ CH_2{-}CH_2 \quad O \end{array} \xleftarrow{-H_2O}$$

b)

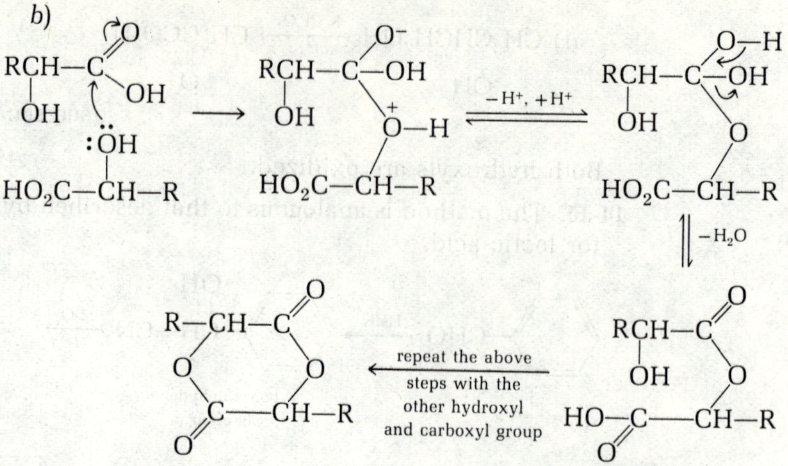

$$RCH-C\underset{OH}{\overset{O}{\diagdown}} \quad \longrightarrow \quad RCH-\underset{OH}{\overset{O^-}{\underset{|}{C}}}-OH \quad \underset{-H^+, +H^+}{\rightleftharpoons} \quad RCH-\underset{OH}{\overset{O-H}{\underset{|}{C}}}-OH$$

$$:\ddot{O}H \qquad\qquad O-H \qquad\qquad O$$

$$HO_2C-CH-R \qquad HO_2C-CH-R \qquad HO_2C-CH-R$$

$$\Big\Downarrow -H_2O$$

$$R-CH-C\overset{O}{\diagdown} \quad \underset{\text{repeat the above}}{\overset{}{\longleftarrow}} \quad RCH-C\overset{O}{\diagdown}$$

$$\underset{O}{\overset{|}{O}} \qquad\qquad \underset{\text{steps with the}}{\underset{\text{other hydroxyl}}{\underset{\text{and carboxyl group}}{}}} \qquad OH \qquad O$$

$$C-CH-R \qquad\qquad HO-C-CH-R$$

$$\overset{}{O} \qquad\qquad\qquad O$$

c)

$$CH_3CHCH_2CO_2H \quad \underset{}{\overset{+H^+}{\rightleftharpoons}} \quad CH_3-CH-\overset{H}{\underset{}{C}}-CO_2H$$

$$:\ddot{O}H \qquad\qquad\qquad :\overset{+}{O}\quad H$$

$$\qquad\qquad\qquad\qquad H \quad H$$

$$CH_3CH=CHCO_2H \quad \underset{-H^+}{\overset{-H_2O,}{\longleftarrow}}$$

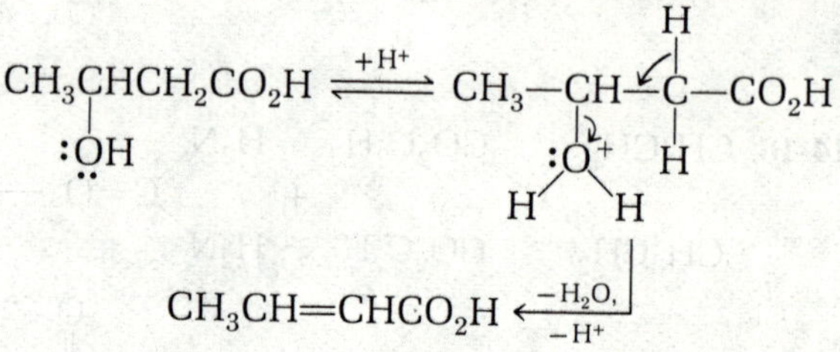

d)

$$\text{(structures shown)}$$

$$\Big\Updownarrow \begin{array}{c} -H^+, \\ +H^+ \end{array}$$

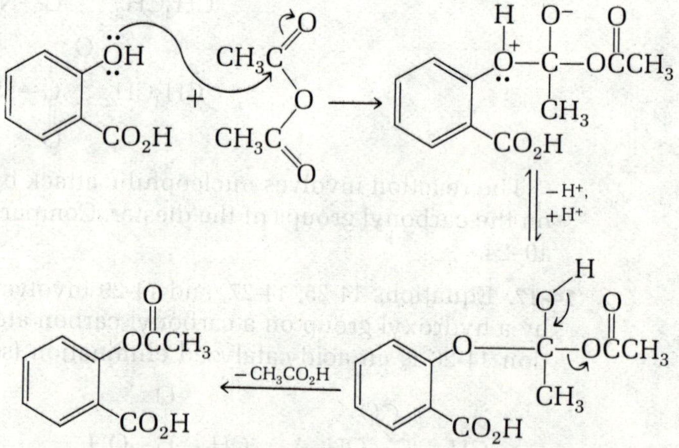

14-18. The mechanism is analogous to that shown in equation 14-10.

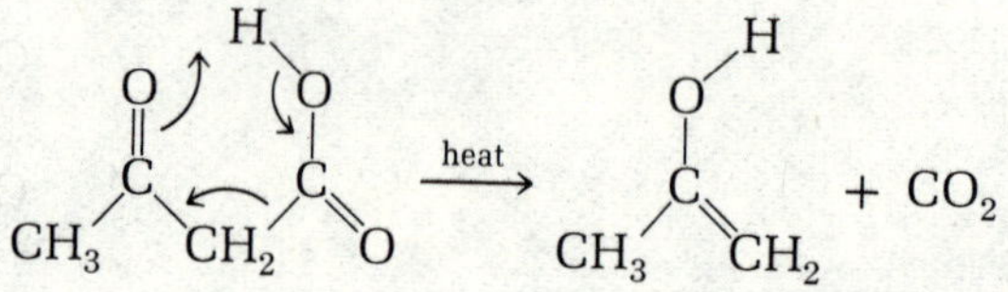

$$CH_3-\overset{O}{\underset{}{C}}-CH_2-\overset{O}{\underset{}{C}}=O \quad \overset{heat}{\longrightarrow} \quad CH_3-\overset{OH}{\underset{}{C}}=CH_2 \quad + \quad CO_2$$

The product is the enol form of acetone, which then ketonizes.

$$CH_3\underset{OH}{\overset{H}{C}}=CH_2 \underset{-H^+}{\rightleftharpoons} \left[CH_3\underset{O^-}{C}=CH_2 \longleftrightarrow CH_3\overset{O}{\underset{\parallel}{C}}CH_2^- \right] \underset{+H^+}{\rightleftharpoons} CH_3\overset{O}{\underset{\parallel}{C}}CH_3$$

14-19. The essential equations are 14-32 through 14-36.

a) $CH_3\overset{O}{\underset{\parallel}{C}}CH_2CO_2C_2H_5 + Na^{+-}OC_2H_5 \xrightleftharpoons[\text{(eq 14-33)}]{}$

$$CH_3\overset{O}{\underset{\parallel}{C}}-\overset{-}{C}H-CO_2C_2H_5 + C_2H_5OH$$
$$Na^+$$

$CH_3CH_2Br + CH_3\overset{O}{\underset{\parallel}{C}}\overset{-}{C}HCO_2C_2H_5 \xrightarrow[\substack{\text{(eq 14-34,}\\ \text{first step)}}]{}$
$$\qquad\qquad\qquad\qquad Na^+$$

$$CH_3\overset{O}{\underset{\parallel}{C}}CH-CO_2C_2H_5 + Na^+Br^-$$
$$\underset{CH_2CH_3}{|}$$

$CH_3\overset{O}{\underset{\parallel}{C}}CHCO_2C_2H_5 + Na^+OH^- \longrightarrow CH_3\overset{O}{\underset{\parallel}{C}}CHCO_2^-Na^+ + C_2H_5OH$
$$\underset{CH_2CH_3}{|} \qquad\qquad\qquad\qquad\quad \underset{CH_2CH_3}{|}$$

$$\big\Updownarrow H^+$$

$CH_3\overset{O}{\underset{\parallel}{C}}CH_2CH_2CH_3 \xleftarrow[\substack{-CO_2 \\ \text{(eq 14-32)}}]{\text{heat}} CH_3\overset{O}{\underset{\parallel}{C}}CHCO_2H + Na^+$
$$\qquad\qquad\qquad\qquad\qquad\qquad\qquad\quad \underset{CH_2CH_3}{|}$$

b) $CH_3\overset{O}{\underset{\parallel}{C}}-CH-CH_2-\bigcirc$
$$\underset{CH_3}{|}$$

These two R groups must be
introduced, as in eq 14-34.

The first step is the same as in part *a*. Then

$$\text{Na}^+\text{-OC}_2\text{H}_5$$

1. NaOH
2. acidify

c) The starting material is a β-keto ester and reacts in a manner analogous to acetoacetic ester.

14-20. Follow the pattern shown in equations 14-37a,b,c.

$$\text{CH}_3\text{CH}_2\text{CO}_2\text{C}_2\text{H}_5 + \text{Na}^+\text{-OC}_2\text{H}_5 \rightleftharpoons \text{CH}_3\bar{\text{C}}\text{HCO}_2\text{C}_2\text{H}_5 + \text{C}_2\text{H}_5\text{OH}$$
$$\text{Na}^+$$

$$\underset{O}{\overset{O}{\parallel}}\quad\quad\quad Na^+$$

$$CH_3CH_2\overset{O}{\overset{\parallel}{C}}OC_2H_5 + \overset{-}{C}HCO_2C_2H_5 \rightleftharpoons CH_3CH_2\overset{O^-Na^+}{\overset{|}{\underset{|}{C}}}—CHCO_2C_2H_5$$

$$\underset{CH_3}{} \quad\quad\quad \underset{OC_2H_5\;\;CH_3}{}$$

$$CH_3CH_2\overset{O}{\overset{\parallel}{C}}—CHCO_2C_2H_5 \rightleftharpoons$$

$$\underset{CH_3}{}$$

The product reacts immediately with the Na$^+$–OC$_2$H$_5$ present to form the enolate salt, which is converted to the desired ketoester by acidification at the end of the reaction.

$$CH_3CH_2\overset{O}{\overset{\parallel}{C}}—\overset{Na^+}{\overset{|}{\underset{|}{\overset{-}{C}}}}—CO_2C_2H_5 \xrightarrow{H^+} CH_3CH_2\overset{O}{\overset{\parallel}{C}}CHCO_2C_2H_5$$

$$\underset{CH_3}{} \quad\quad\quad\quad\quad\quad\quad \underset{CH_3}{}$$

14-21. *a*) The enolate anion must be formed from the ethyl acetate, since ethyl formate has no α hydrogens. The key step is then

$$\overset{O}{\overset{\parallel}{HC}}OC_2H_5 + \overset{-}{C}H_2CO_2C_2H_5 \longrightarrow H—\overset{O^-}{\overset{|}{\underset{|}{C}}}—CH_2CO_2C_2H_5$$

ethyl formate $\quad\quad\quad\quad\quad\quad\quad\quad\quad\quad \underset{OC_2H_5}{}$

$$\Big\downarrow ^{-OC_2H_5}$$

$$\overset{O}{\overset{\parallel}{HC}}CH_2CO_2C_2H_5$$

This is the cross-Claisen product.

b)

$$\begin{matrix} CH_2—CO_2C_2H_5 \\ CH_2 \\ | \\ CH_2 \\ CH_2—CO_2C_2H_5 \end{matrix} \xrightarrow{Na^+-OC_2H_5} \begin{matrix} \overset{-}{C}H—CO_2C_2H_5 \\ CH_2 \\ | \\ CH_2 \\ CH_2—C—OC_2H_5 \\ O \end{matrix}$$

$$\Big\downarrow$$

$$\begin{matrix} CH_2 \\ CH_2 \\ CH_2 \\ C \\ O \end{matrix} CHCO_2C_2H_5 \xleftarrow{-OC_2H_5} \begin{matrix} CH_2 \\ CH_2 \\ CH_2 \\ C \\ ^{-}O \quad OC_2H_5 \end{matrix} CH—CO_2C_2H_5$$

This is the way the starting material in problem 14-19c is usually prepared.

14-22. The equations with ethyl butyrate and ethyl isobutyrate are exactly analogous to those written out in detail in the answer to problem 14-20. When ethoxide ion is eliminated in the penultimate stage of the reaction, the products will be

$$CH_3CH_2CH_2\overset{\overset{\displaystyle O}{\|}}{C}\underset{\underset{\displaystyle CH_2CH_3}{|}}{CH}CO_2C_2H_5 \quad \text{and} \quad \underset{CH_3}{\overset{CH_3}{>}}CH-\overset{\overset{\displaystyle O}{\|}}{C}-\underset{\underset{\displaystyle CH_2CH_3}{|}}{\overset{\overset{\displaystyle CH_3}{|}}{C}}-CO_2C_2H_5$$

In the case of ethyl butyrate, the reversible steps can be driven to completion by forming the enolate anion of the product:

$$CH_3CH_2CH_2\overset{\overset{\displaystyle O}{\|}}{C}\underset{\underset{\displaystyle CH_2CH_3}{|}}{CH}CO_2C_2H_5 + Na^{+-}OC_2H_5 \longrightarrow$$

$$CH_3CH_2CH_2\overset{\overset{\displaystyle O}{\|}\,Na^+}{C}\underset{\underset{\displaystyle CH_2CH_3}{|}}{\overset{-}{C}}CO_2C_2H_5 + C_2H_5OH$$

But this is not possible for the ethyl isobutyrate product, which does not posess an acidic hydrogen.

$$\underset{CH_3}{\overset{CH_3}{>}}CH-\overset{\overset{\displaystyle O}{\|}}{C}-\underset{\underset{\displaystyle CH_2CH_3}{|}}{\overset{\overset{\displaystyle CH_3}{|}}{C}}-CO_2C_2H_5 \quad \text{— no H on this carbon atom}$$

Therefore all the reaction steps are reversible, and the condensation does not go well unless special conditions are used.

14-23. The mechanism is that of hemiacetal formation (eq 9-22).

14-24. The substance is an acid, so the formula $C_4H_8O_3$ can therefore be represented as $C_3H_7O-COOH$. The reaction with acetyl chloride suggests that the remaining oxygen is present as a hydroxyl group, since alcohols are known to react with acetyl chloride to form esters. The fact that the acid loses water on heating ($C_4H_8O_3 \longrightarrow C_4H_6O_2$) tends to support the presence of an alcohol function, with a hydrogen on the adjacent carbon(s). The positive iodoform test suggests that the alcohol must be present as CH_3CH- (sec 9.7b).

$$\overset{\displaystyle CH_3CH-}{\underset{\displaystyle OH}{}}$$

A satisfactory structure is

$$CH_3CHCH_2CO_2H \xrightarrow{\text{heat}}$$

$$\overset{CH_3}{\underset{H}{}}C=C\overset{CO_2H}{\underset{H}{}}$$

$$\overset{CH_3}{\underset{H}{}}C=C\overset{H}{\underset{CO_2H}{}}$$

$$CH_2=CHCH_2CO_2H$$

Equations for the tests are

$$CH_3CHCH_2CO_2H + CH_3\overset{O}{\overset{\|}{C}}Cl \longrightarrow CH_3CHCH_2CO_2H + HCl$$

$$CH_3CHCH_2CO_2H + 4\,I_2 + 7\,Na^+OH^- \longrightarrow$$

$$CHI_3 + 5\,Na^+I^- + Na^{+-}O_2CCH_2CO_2{}^-Na^+ + 6\,H_2O$$

14-25. Since salicylic acid, $\overset{CO_2H}{\underset{OH}{\bigcirc}}$, is an oxidation product,

(A) must have a formula

$$\overset{C(C_2H_3O_2)}{\underset{OH}{\bigcirc}}$$

Since the side chain includes a carboxyl group (the compound is said to be a phenolic *acid*), the structure can be further refined to

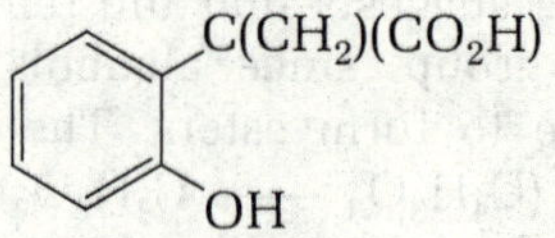

$$C(CH_2)(CO_2H)$$

on benzene ring with OH

The two possibilities are

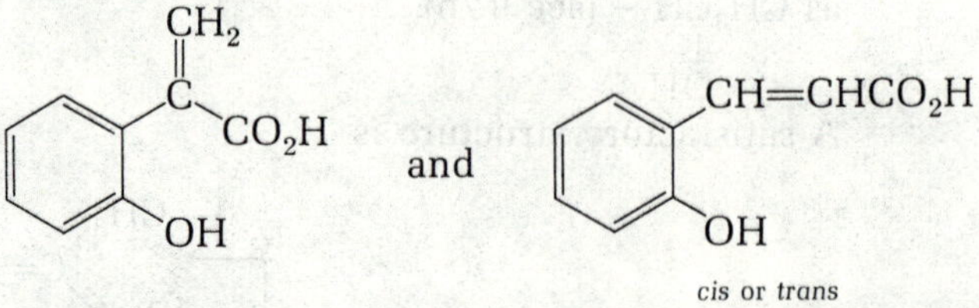

and

cis or trans

Only the second of these will give oxalic acid on oxidation:

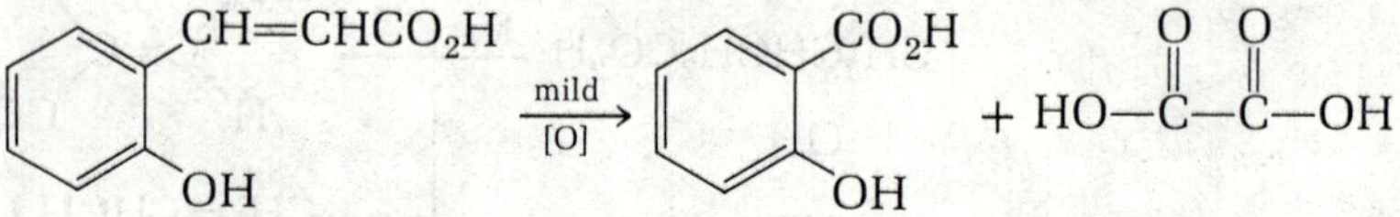

$$\xrightarrow[\text{[O]}]{\text{mild}}$$

$$+\ \ HO-\overset{\displaystyle O}{\underset{\displaystyle \|}{C}}-\overset{\displaystyle O}{\underset{\displaystyle \|}{C}}-OH$$

Therefore the structures of (A) are the *cis-* and *trans-*o-hydroxycinnamic acids. Of these, the *cis* isomer easily loses water to form a lactone; the *trans* cannot, because the carboxyl and hydroxyl groups are too far apart.

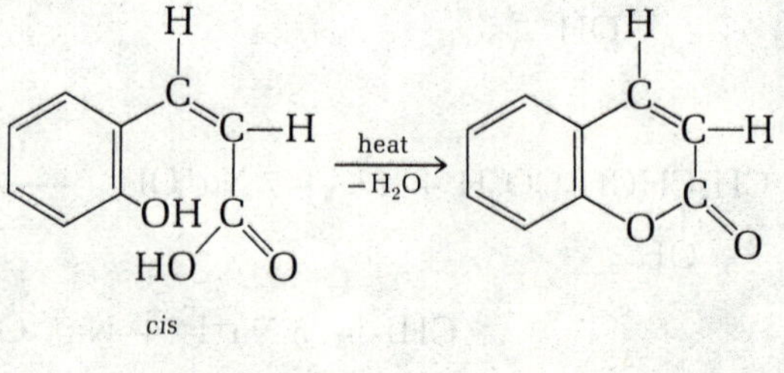

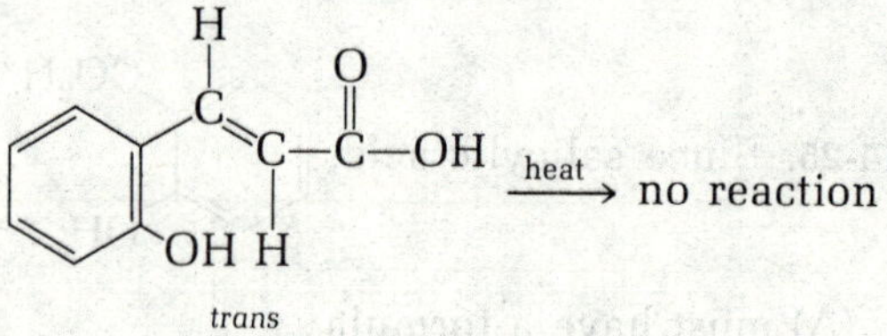

$$\xrightarrow[\text{--H}_2\text{O}]{\text{heat}}$$

cis

$$\xrightarrow{\text{heat}}\ \text{no reaction}$$

trans

CHAPTER FIFTEEN

OPTICAL ISOMERISM

15-1. Each of these definitions can be found explicitly or implicitly in the text. The sections where they appear are indicated below:

a) sec 15.1   g) sec 15.5

b) sec 15.1a   h) sec 15.7

c) sec 15.1a   i) sec 15.9

d) sec 15.1b   j) sec 15.8

e) sec 15.2   k) sec 15.4

f) sec 15.3   l) sec 15.1a

15-2. The molecule has *five* planes of symmetry, each perpendicular to the plane of the molecule, passing through one of the five carbons and bisecting the bond "opposite" the carbon through which it passes. If planar, the molecule also has a plane of symmetry which is the molecular plane (i.e., a plane which passes through all five carbon atoms.

The five twofold axes are in the plane of the molecule; they pass through one carbon atom and bisect the bond on the "opposite" side of the ring.

15-3. Unlike the helix in Figure 15-3, which has a twofold symmetry axis, this helix has no symmetry element, and is dissymmetric. (The axis which goes down the middle of the helix is only an identity axis; i.e., the helix is converted only into itself by a full 360° rotation about this axis—this is therefore not considered a symmetry axis, since every object has an infinite number of such axes).

15-4. a) If a molecule and its mirror image are not superimposable, then the molecule which corresponds to the structure of the mirror image will be an enantiomer of the original molecule.

b) If a molecule has only one asymmetric carbon atom, this will be sufficient to ensure that it has an enantiomer (but it is not necessary that a molecule have an asymmetric carbon to have an enantiomer; see sec 15.10).

c) Most commonly, examine the structure for a plane of symmetry—if none is present, enantiomers are usually possible. But the decisive test is to compare the molecule with its mirror image and determine whether or not they are superimposable.

15-5. a) CH_3 ⓒ $H(Br)CH_2CH_3$

b) CH_3 ⓒ $H(Cl)$ ⓒ $H(Cl)CH_3$

c) C_6H_5 ⓒ $H(OH)CO_2H$

d) $CH_2(OH)$ ⓒ $H(OH)$ ⓒ $H(OH)CH_2OH$

e) ⓒ $H(OH)CH_3$

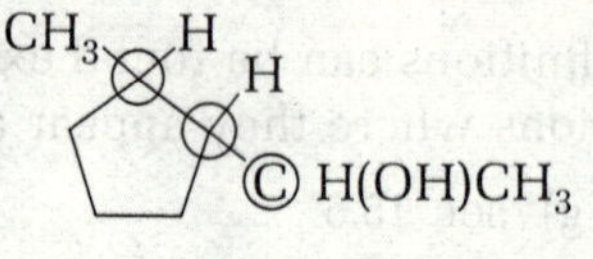

f) This structure has no asymmetric carbon atoms.

15-6. a) $CH_2\overset{*}{C}HCH_3$ with Cl and OH

The carbon marked with an asterisk is asymmetric; the molecule can be optically active.

b)

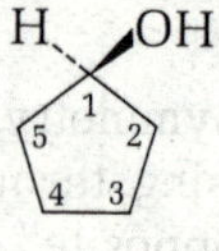

The substance is optically inactive; it has a plane of symmetry perpendicular to the five-membered ring, through carbon 1 and bisecting the C3—C4 bond.

c) CH_2CH_2OH with Br

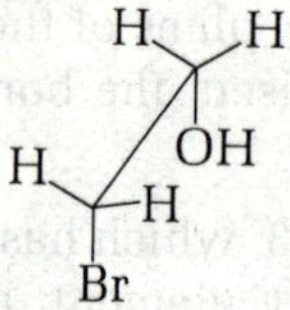

Optically inactive; no asymmetric carbons and, in the eclipsed conformation shown, a plane of symmetry through the two carbons, the Br and OH.

d) $-\overset{*}{C}HCH_3$ with Cl

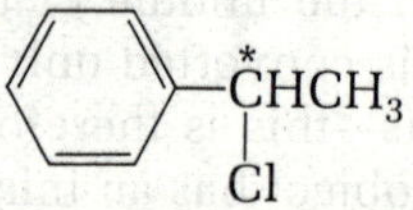

Optically active; the asterisked carbon is asymmetric.

e) $CH_3\overset{*}{C}HCO_2H$ with NH_2

Optically active; the asterisked carbon is asymmetric.

f)

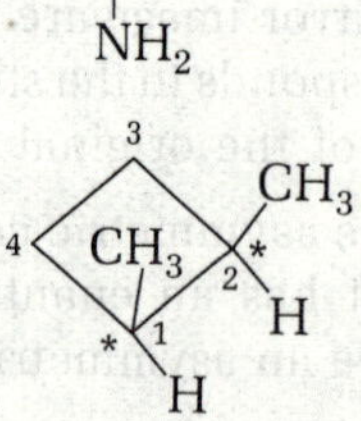

Optically inactive; although the two asterisked carbons are asymmetric, they are also identical. The molecule has a plane of symmetry perpendicular to the plane of the four-membered ring and bisecting the C1—C2 and C3—C4 bonds.

g)

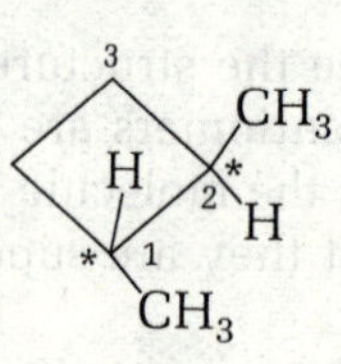

Optically active; no plane or center of symmetry in the molecule.

h)

Br   Br
  \   /
   C
  /   \
CH$_3$ CH$_3$

Optically inactive; the molecule has several planes of symmetry, for example the one which passes through the three carbon atoms.

15-7.

CO$_2$H   CO$_2$H   CO$_2$H   CO$_2$H

H—OH $\xrightarrow{\ \sigma\ }$ H—CH$_3$ $\xrightarrow{\ \sigma\ }$ HO—CH$_3$ $\xrightarrow{\ \sigma\ }$ HO—H

CH$_3$ OH   H   CH$_3$

(−)   (+)   (−)   (+)

The symbol $\xrightarrow{\ \sigma\ }$ represents an inversion of configuration. Note that at least three steps are necessary to get from the (−) structure to the particular (+) Fischer projection shown in the problem.

15-8. Start with the Fischer projection of the (−) enantiomer as it is usually written:

CO$_2$H   CO$_2$H   OH   H

H—OH $\xrightarrow{\ \sigma\ }$ CH$_3$—OH $\xrightarrow{\ \sigma\ }$ CH$_3$—CO$_2$H $\xrightarrow{\ \sigma\ }$ CH$_3$—CO$_2$H

CH$_3$ H   H   OH

(−)   (+)   (−)   (+)

An odd number (at least three) of interchanges is necessary.

15-9. In each case, determine whether an even or odd number of group interchanges are necessary to convert the projection shown to the usual projection of the (−) enantiomer.

a)

CH$_3$ CH$_3$ CO$_2$H

HO—H $\xrightarrow{\ \sigma\ }$ H—OH $\xrightarrow{\ \sigma\ }$ H—OH

CO$_2$H   CO$_2$H   CH$_3$

(−)   (+)   (−)

Since an even number of interchanges is involved, the original projection must be of the (−) isomer.

b)

H   CO$_2$H   CO$_2$H

HO$_2$C—CH$_3$ $\xrightarrow{\ \sigma\ }$ H—CH$_3$ $\xrightarrow{\ \sigma\ }$ H—OH

OH   OH   CH$_3$

(−)   (+)   (−)

c)

OH   CO$_2$H   CO$_2$H

HO$_2$C—H $\xrightarrow{\ \sigma\ }$ HO—H $\xrightarrow{\ \sigma\ }$ H—OH

CH$_3$ CH$_3$ CH$_3$

(−)   (+)   (−)

d)

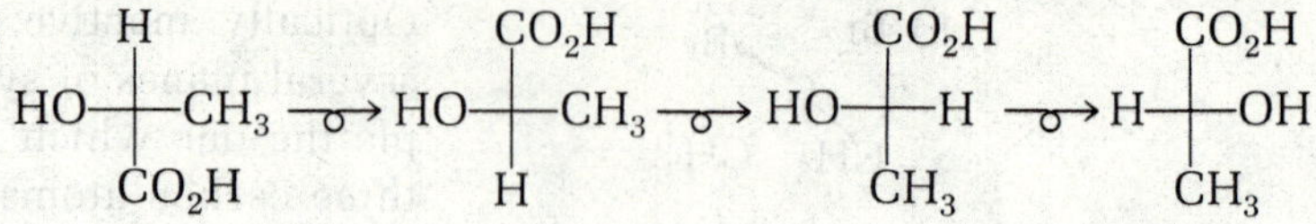

$$HO-\overset{H}{\underset{CO_2H}{|}}-CH_3 \longrightarrow HO-\overset{CO_2H}{\underset{H}{|}}-CH_3 \longrightarrow HO-\overset{CO_2H}{\underset{CH_3}{|}}-H \longrightarrow H-\overset{CO_2H}{\underset{CH_3}{|}}-OH$$

$$(+) \qquad\qquad (-) \qquad\qquad (+) \qquad\qquad (-)$$

15-10 The rules are given in section 15.6.

a) $OH > CH_3CH_2 > CH_3 > H$ b) $Cl > C_6H_5 > CH_3 > H$

C_6H_5 takes precedence over CH_3, because the carbon marked with an asterisk in C_6H_5-

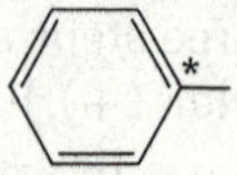

has three bonds to other carbons, whereas in CH_3 those bonds are to hydrogen (which has a lower atomic number than carbon).

c) $-OH > -CH_2Cl > -CH_2OH > -CH_3$

d) $-C(CH_3)_3 > -CH(CH_3)_2 > -CH_2CH_3 > -CH_3$

15-11. Draw a representation of the tetrahedron, and locate the lowest priority group; then look at the model from the opposite side of that group and locate the remaining three groups in a clockwise order according to their priority.

a)

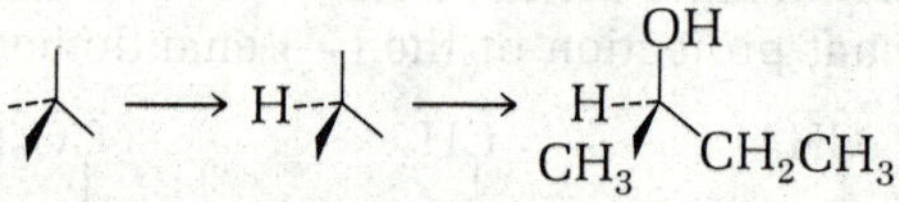

Other equivalent structures are

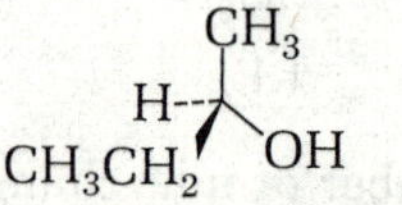

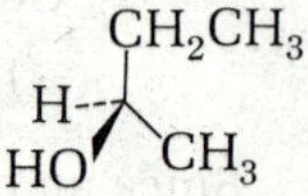

All have the **R** configuration.

b)

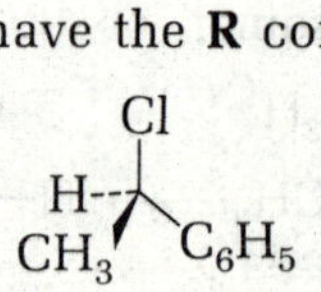

c)

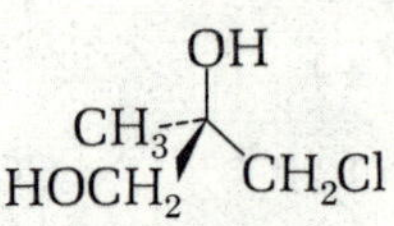

d)

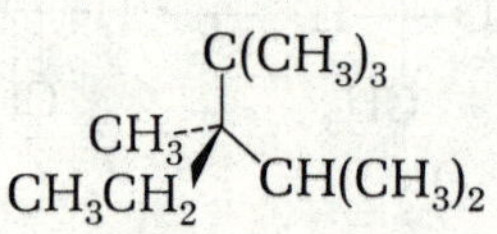

15-12. *a)*

$$CH_3CH_2 \overset{\displaystyle CH_3}{\underset{\displaystyle H}{\rule[0.5ex]{2em}{0.4pt}}} OH$$

c)

$$ClCH_2 \overset{\displaystyle CH_2OH}{\underset{\displaystyle CH_3}{\rule[0.5ex]{2em}{0.4pt}}} OH$$

b)

$$C_6H_5 \overset{\displaystyle CH_3}{\underset{\displaystyle H}{\rule[0.5ex]{2em}{0.4pt}}} Cl$$

d)

$$(CH_3)_2CH \overset{\displaystyle CH_2CH_3}{\underset{\displaystyle CH_3}{\rule[0.5ex]{2em}{0.4pt}}} C(CH_3)_3$$

In each case, notice that the remaining three groups, in priority order, are arranged in a clockwise manner.

15-13. The following are examples; there may be other possibilities.

a) $CH_3\overset{*}{C}HCH_2CH_3$
  $\quad\ \ \underset{\displaystyle OH}{|}$

c) $HOCH_2\overset{*}{C}HCH_2CH_3$
  $\qquad\quad\ \underset{\displaystyle OH}{|}$

b) $CH_3\overset{*}{C}HCH_2CH_2CH_3$
  $\quad\ \ \underset{\displaystyle Br}{|}$

d) $CH_3\overset{*}{C}HCH{=}CH_2$
  $\quad\ \ \underset{\displaystyle CH_2CH_3}{|}$

In each case, the asymmetric carbon atom is marked with an asterisk.

15-14.

$$CH_2{-}O{-}\overset{\displaystyle O}{\overset{\|}{C}}{-}(CH_2)_{16}CH_3$$
$$\underset{\displaystyle |}{} \qquad O$$
$$H{-}\overset{*}{C}{-}O{-}\overset{\|}{C}{-}(CH_2)_{14}CH_3$$
$$CH_2{-}O{-}\overset{\displaystyle |}{\underset{\displaystyle O}{\overset{\|}{C}}}{-}(CH_2)_7CH{=}CH(CH_2)_7CH_3$$

The carbon marked with an asterisk is asymmetric; the fat might, therefore, be optically active

Catalytic hydrogenation of the double bond would convert the oleate group to a stearate. In this case the upper and lower groups would become identical ($-CH_2O\overset{\displaystyle O}{\overset{\|}{C}}(CH_2)_{16}CH_3$) and the compound would no longer have an asymmetric carbon or be capable of optical activity.

15-15. Consider first the "rear" asymmetric carbon of the upper left formula in Figure 15-20. The priority order will be

$$-\overset{\displaystyle O}{\overset{\|}{C}}{-}OH > -\overset{\displaystyle OH}{\underset{\displaystyle CO_2H}{\overset{\displaystyle |}{C}}}{-}H > -CH_3 > H$$

The hydrogen has the lowest atomic number of the four atoms attached to the asymmetric carbon, and clearly has lowest priority. The remaining three groups have the same atomic number (all carbon) at the first atom of attachment; one must therefore go out to the second atoms.

$$
\underset{\substack{\text{three bonds}\\ \text{to oxygens}}}{-\overset{\displaystyle O}{\overset{\|}{C}}-O} \qquad \underset{\substack{\text{only one bond to oxygen,}\\ \text{the others to atoms of}\\ \text{lower atomic number}}}{-\overset{\displaystyle O}{\overset{\|}{\underset{\underset{C}{|}}{C}}}-H} \qquad \underset{\substack{\text{all bonds to}\\ \text{hydrogen}}}{-\overset{\displaystyle H}{\underset{\underset{H}{|}}{\overset{|}{C}}}-H}
$$

If the model is viewed from the face "opposite" the lowest priority group, the priority order of the remaining groups is achieved by a clockwise or **R** motion.

$$
\begin{array}{c}
\text{CH}_3 \diagup \quad \diagdown \text{H} \\
\text{HO} \diagup \diagdown \text{H}
\end{array}
\quad
\begin{array}{c}
\text{CO}_2\text{H} \\
| \\
\text{CO}_2\text{H}
\end{array}
$$

Consider now the "front" asymmetric carbon atom in the same molecule. The priority order will be

$$
\text{OH} > \text{CO}_2\text{H} > -\underset{\underset{\text{CH}_3}{|}}{\overset{\overset{\text{CO}_2\text{H}}{|}}{C}}-\text{H} > \text{H}
$$

The oxygen (OH) has the highest atomic number and the hydrogen (H) the lowest, so these groups clearly belong at opposite ends of the priority order. One must go to the next atoms to decide between the other two groups:

$$
\underset{\substack{\text{three bonds}\\ \text{to oxygens}}}{-\overset{\displaystyle O}{\overset{\|}{C}}-O} \qquad\qquad \underset{\substack{\text{bonds to carbon and hydrogen,}\\ \text{both lower priority order than oxygen.}}}{-\overset{\overset{\text{C}}{|}}{\underset{\underset{C}{|}}{C}}-H}
$$

Once again, the configuration is **R**.

The remaining assignments in Figure 15-20 are automatic, the configurations being either identical with or mirror images of those in the first formula.

Consider now the "rear" asymmetric carbon in the upper left formula in Figure 15-22. The priority order is

$$
\text{OH} > \text{CO}_2\text{H} > -\underset{\underset{\text{H}}{|}}{\overset{\overset{\text{OH}}{|}}{C}}-\text{CO}_2\text{H} > \text{H}
$$

for reasons completely analogous to those given above. The configuration is **R**.

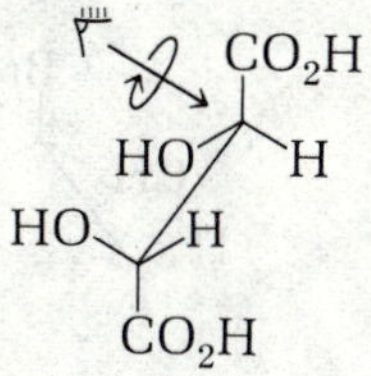

The priority order for the "front" asymmetric carbon is the same, and if viewed from the face "opposite" the hydrogen, it is seen that the configuration here too is **R**.

The remaining assignments in Figure 15-22 are automatic, the configurations being either identical with or mirror images of those in the first formula.

15-16. For three different asymmetric carbons, $2^3 = 8$. The possibilities are

R—R—R	S—R—R
R—R—S	S—R—S
R—S—R	S—S—R
R—S—S	S—S—S

For four different asymmetric carbons, $2^4 = 16$. The possibilities are

R—R—R—R	R—S—R—R	S—R—R—R	S—S—R—R
R—R—R—S	R—S—R—S	S—R—R—S	S—S—R—S
R—R—S—R	R—S—S—R	S—R—S—R	S—S—S—R
R—R—S—S	R—S—S—S	S—R—S—S	S—S—S—S

15-17. The ordinary formula is

$$CH_3 - \overset{*}{C}H - \overset{*}{C}H - CH_3$$
$$\qquad\quad | \qquad\; |$$
$$\qquad\quad Br \qquad Br$$

The two asymmetric carbons are marked with an asterisk. The priority order at each will be

$$Br > -CH(Br)CH_3 > CH_3 > H$$

Draw a "sawhorse" and pick one position for the hydrogen.

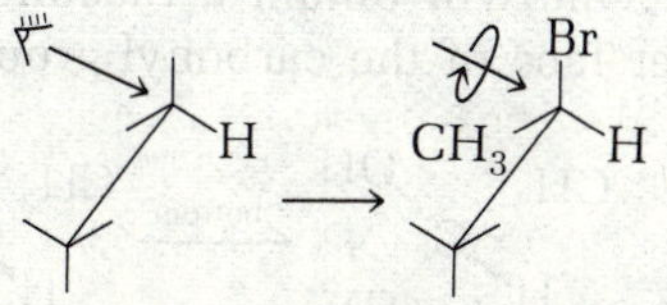

Since the position of the —CH(Br)CH$_3$ group is fixed, one need simply locate the remaining two groups, as shown above.

Now locate the hydrogen anywhere on the front carbon and locate the remaining groups similarly.

A Fischer projection formula for this molecule is

$$
\begin{array}{c}
CH_3 \\
Br \!-\!\!\!-\! H \\
H \!-\!\!\!-\! Br \\
CH_3
\end{array}
$$

Its enantiomer will be
$$
\begin{array}{c}
CH_3 \\
H \!-\!\!\!-\! Br \\
Br \!-\!\!\!-\! H \\
CH_3
\end{array}
$$
; its diastereomer is
$$
\begin{array}{c}
CH_3 \\
H \!-\!\!\!-\! Br \\
H \!-\!\!\!-\! Br \\
CH_3
\end{array}
$$

The latter structure clearly has a plane of symmetry and is a meso structure (optically inactive).

15-18. *a)* Since both reactants are optically inactive, the product—2,3-dichlorobutane—will also be inactive. It will be a mixture of racemic (±) and meso isomers, whose exact composition will depend on the *cis/trans* ratio of the starting 2-butene.

b) This is an S_N2 reaction which should occur with complete inversion of configuration (sec 8.4). The product, $C_6H_5CH(CN)CH_3$, should therefore be optically active and have the opposite configuration of the starting material.

c) Reduction gives $CH_3CH_2\!-\!\underset{\underset{OH}{|}}{CH}\!-\!CH_2CH_3$, which no longer has an asymmetric carbon atom. The product is optically inactive.

d) One will obtain a racemic mixture, since CN^- will attack either face of the carbonyl group:

15-19.

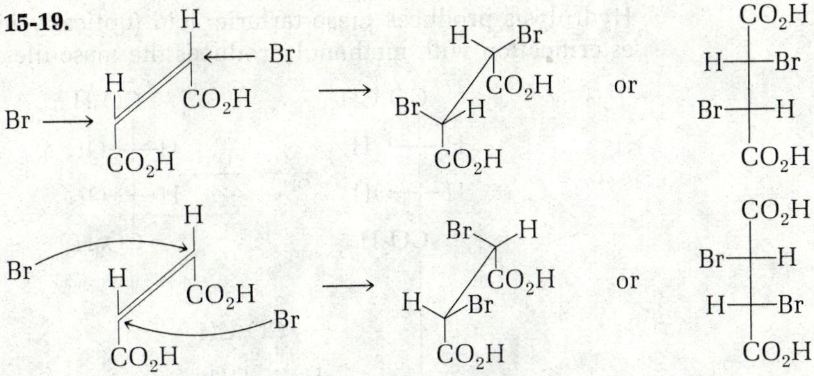

"*Trans*" addition of Br_2 to maleic acid gives equal amounts of the two enantiomeric dibromosuccinic acids which can be resolved.

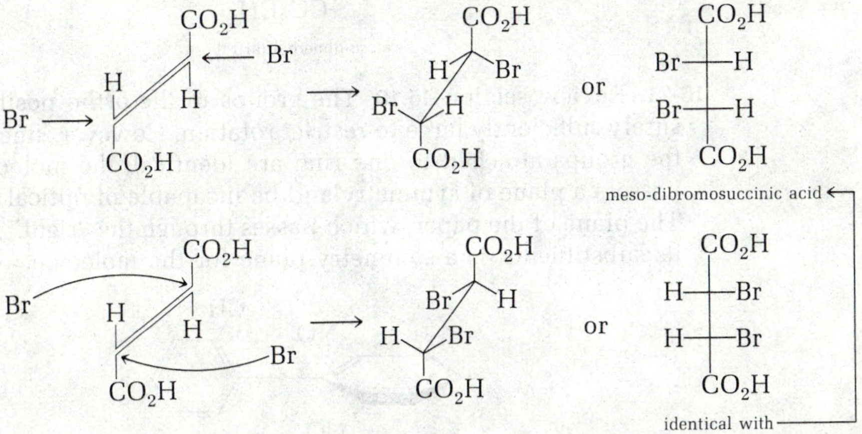

meso-dibromosuccinic acid

identical with

"*Trans*" addition of Br_2 to fumaric acid gives meso-dibromosuccinic acid which cannot be resolved. This result proves that the two bromines add from opposite "sides" of the double bond. If bromine addition were "*cis*" (both Br atoms from the same side of the double bond), then the result would have been reversed; maleic acid would have given meso-dibromosuccinic acid, and fumaric acid would have given a resolvable product.

15-20. The compound must be the monomethyl ester of meso-tartaric acid.

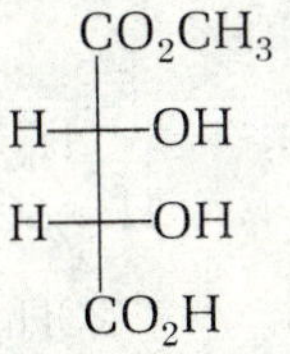

The two asymmetric carbons are different and the compound can be optically active.

Hydrolysis produces meso-tartaric acid (optically inactive), and esterification with methanol produces the meso-diester.

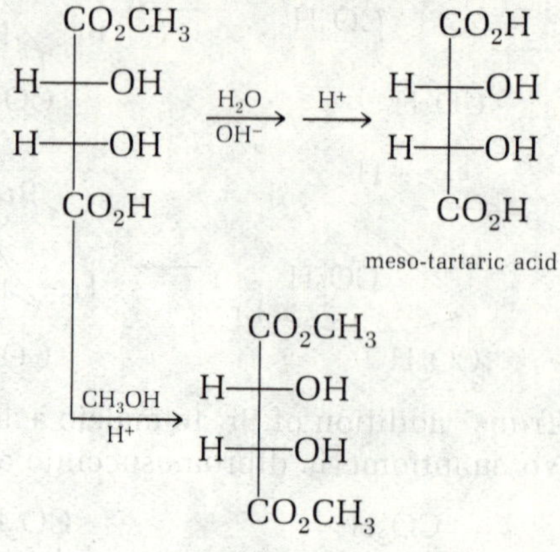
CO2CH3
H—OH
H—OH
CO2H
H2O
OH−
H+
CO2H
H—OH
H—OH
CO2H
CH3OH
H+
CO2CH3
H—OH
H—OH
CO2CH3

meso-tartaric acid

meso-dimethyl tartrate

15-21. Review section 15.10. The groups in the *ortho* positions are surely sufficiently large to restrict rotation. However, since two of the groups attached to one ring are identical, the molecule will possess a plane of symmetry and be incapable of optical activity. The plane of the paper, which passes through the "right" ring and its substituents, is a symmetry plane for the molecule.

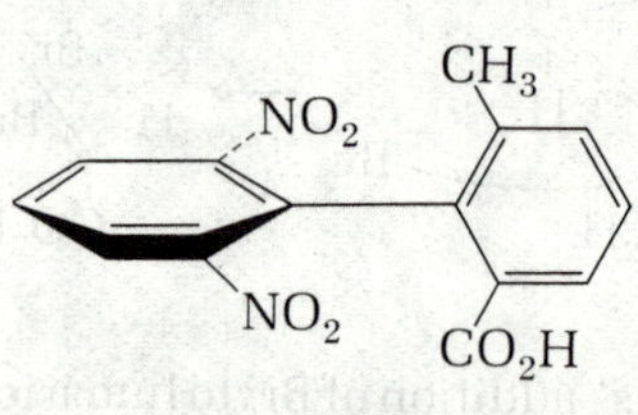
CH3
NO2
NO2
CO2H

15-22. The second and fourth conformations (numbering left to right) have no symmetry element and are chiral. The first conformation has a symmetry plane which is perpendicular to and bisects the C2—C3 bond and is achiral. The midpoint of the C2—C3 bond in the third structure is a center of symmetry; this structure too is achiral.

15-23. Suppose we consider the "sawhorse" formula for one of the two enantiomers.

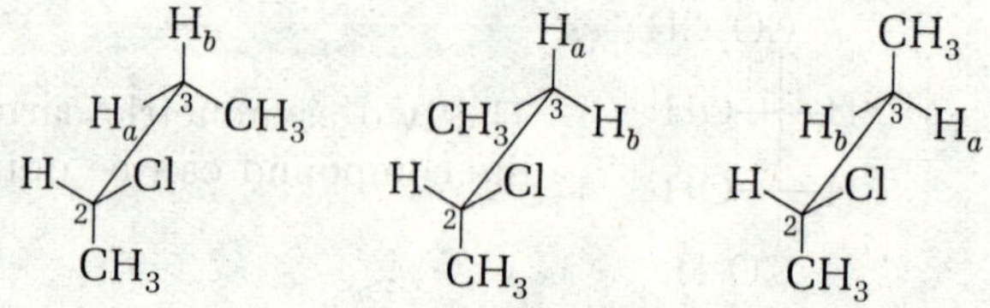
Hb
Ha
3
CH3
H
Cl
2
CH3
Ha
CH3
3
Hb
H
Cl
2
CH3
CH3
Hb
3
Ha
H
Cl
2
CH3

No matter which conformer we choose, the two hydrogens H_a and H_b (one of which is replaced by chlorine when the compound is chlorinated to give 2,3-dichlorobutane) are in different environments. For example, in the "left" conformer, H_a is flanked by H and CH_3 on C2, whereas H_b is flanked by H and Cl. If we rotate to get the "right" conformer, H_b now has the position occupied formerly by H_a (flanked by H and CH_3) *but H_a does not have the position occupied formerly by H_b* (it is flanked by CH_3 and Cl not H and Cl). The positions of H_a and H_b cannot be interchanged; these hydrogens are said to be *diastereotopic,* since if one or the other were replaced by some group which would make C3 asymmetric, the products would be diastereomers.

Since H_a and H_b are *not* equivalent, there is no reason why they should react identically. Thus the ratio of meso to racemic product need not be 1:1. The argument applies regardless of which isomer of 2-chlorobutane we consider, and is therefore true of the racemic mixture as well.

15-24.

$$CH_3 \text{---} \underset{\underset{CH_2OH}{|}}{\overset{\overset{CO_2H}{|}}{}} \text{---} CH_2CH_3 \longrightarrow CH_3 \text{---} \underset{\underset{CH_2OH}{|}}{\overset{\overset{CO_2C_2H_5}{|}}{}} \text{---} CH_2CH_3$$

$$\text{A} \qquad\qquad\qquad \text{B}$$

$$\downarrow$$

$$CH_3 \text{---} \underset{\underset{CO_2H}{|}}{\overset{\overset{CH_2OH}{|}}{}} \text{---} CH_2CH_3 \longleftarrow CH_3 \text{---} \underset{\underset{CO_2H}{|}}{\overset{\overset{CO_2C_2H_5}{|}}{}} \text{---} CH_2CH_3$$

$$\text{D} \qquad\qquad\qquad \text{C}$$

A and **D** are clearly enantiomers, since their Fischer projections are interconvertible by one group interchange (or an odd number of them).

This experiment unequivocally establishes that two enantiomers are otherwise identical but have two groups in exchanged positions (the sequence converts the CO_2H in **A** to CH_2OH, and the CH_2OH to CO_2H, without breaking a bond to the asymmetric carbon atom).

15-25. Paper is primarily cellulose, which has many asymmetric carbon atoms (see sec 16.7c). Being asymmetric, cellulose can interact in a diastereomeric way with the two enantiomers of a racemic mixture (sec 15.12). If the interaction is sufficiently different, resolution may occur.

CHAPTER SIXTEEN

CARBOHYDRATES

16-1. These definitions, usually with examples, are given in the following sections of the text:

a) 16.1   f) 16.2b

b) 16.1   g) 16.2b

c) 16.1, 16.2   h) 16.1, 16.2b, 16.2c, 16.5c

d) 16.1, 16.6   i) 16.5b

e) 16.1, 16.7   j) 16.5a

16-2. a)

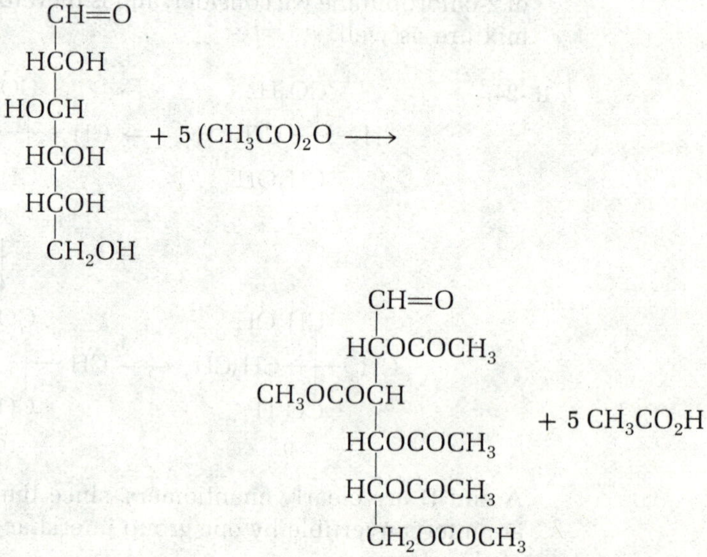

CH=O
|
HCOH
|
HOCH   + 5 (CH$_3$CO)$_2$O ⟶
|
HCOH
|
HCOH
|
CH$_2$OH

CH=O
|
HCOCOCH$_3$
|
CH$_3$OCOCH
| + 5 CH$_3$CO$_2$H
HCOCOCH$_3$
|
HCOCOCH$_3$
|
CH$_2$OCOCH$_3$

glucose pentaacetate

b)

CH=O
|
HCOH
|
HOCH   + Br$_2$ + H$_2$O ⟶
|
HCOH
|
HCOH
|
CH$_2$OH

CO$_2$H
|
HCOH
|
HOCH   + 2 HBr
|
HCOH
|
HCOH
|
CH$_2$OH

gluconic acid

c)

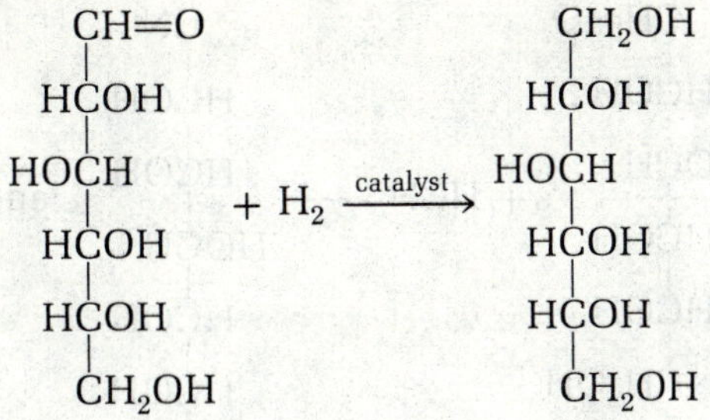

$$
\begin{array}{c}
\text{CH}{=}\text{O} \\
\text{HCOH} \\
\text{HOCH} \\
\text{HCOH} \\
\text{HCOH} \\
\text{CH}_2\text{OH}
\end{array}
\quad + \text{H}_2 \xrightarrow{\text{catalyst}}
\begin{array}{c}
\text{CH}_2\text{OH} \\
\text{HCOH} \\
\text{HOCH} \\
\text{HCOH} \\
\text{HCOH} \\
\text{CH}_2\text{OH}
\end{array}
$$

sorbitol

d)

$$
\begin{array}{c}
\text{CH}{=}\text{O} \\
\text{HCOH} \\
\text{HOCH} \\
\text{HCOH} \\
\text{HCOH} \\
\text{CH}_2\text{OH}
\end{array}
\quad + \text{NH}_2\text{OH} \longrightarrow
\begin{array}{c}
\text{CH}{=}\text{NOH} \\
\text{HCOH} \\
\text{HOCH} \\
\text{HCOH} \\
\text{HCOH} \\
\text{CH}_2\text{OH}
\end{array}
\quad + \text{H}_2\text{O}
$$

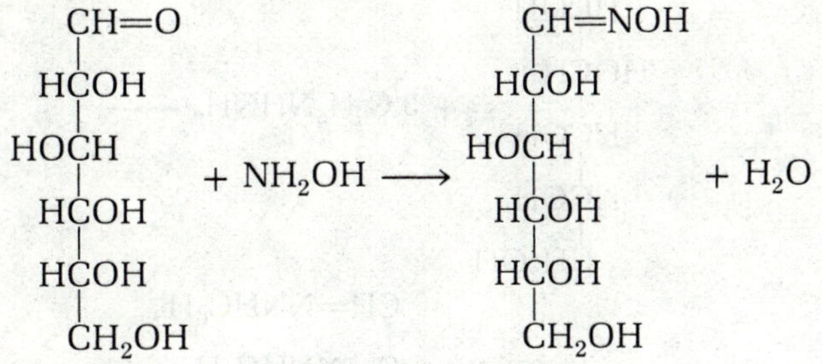

glucose oxime

e)

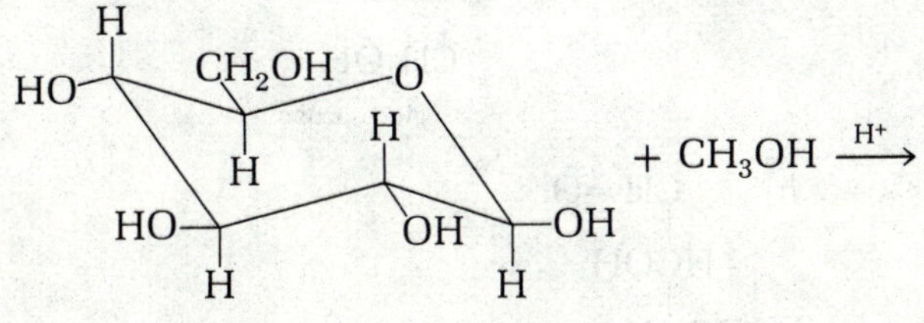

and/or the α-form

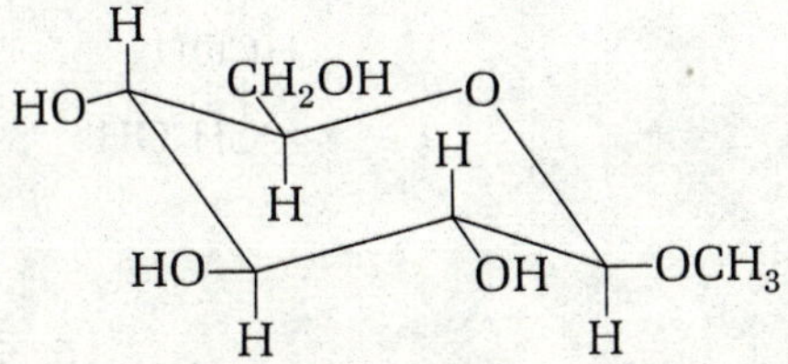

$+ \text{CH}_3\text{OH} \xrightarrow{\text{H}^+}$

methyl β-D-glucoside
(and the α-form)

f)

$$\begin{array}{c} \text{CH}=\text{O} \\ \text{HCOH} \\ \text{HOCH} \\ \text{HCOH} \\ \text{HCOH} \\ \text{CH}_2\text{OH} \end{array} \quad + \text{ HCN} \longrightarrow \quad \begin{array}{c} \text{CN} \\ \text{HCOH} \\ \text{HCOH} \\ \text{HOCH} \\ \text{HCOH} \\ \text{HCOH} \\ \text{CH}_2\text{OH} \end{array} \quad \text{and/or} \quad \begin{array}{c} \text{CN} \\ \text{HOCH} \\ \text{HCOH} \\ \text{HOCH} \\ \text{HCOH} \\ \text{HCOH} \\ \text{CH}_2\text{OH} \end{array}$$

glucose cyanohydrin

g)

$$\begin{array}{c} \text{CH}=\text{O} \\ \text{HCOH} \\ \text{HOCH} \\ \text{HCOH} \\ \text{HCOH} \\ \text{CH}_2\text{OH} \end{array} \quad + \text{ 3 C}_6\text{H}_5\text{NHNH}_2 \longrightarrow$$

$$\begin{array}{c} \text{CH}=\text{NNHC}_6\text{H}_5 \\ \text{C}=\text{NNHC}_6\text{H}_5 \\ \text{HOCH} \\ \text{HCOH} \\ \text{HCOH} \\ \text{CH}_2\text{OH} \end{array} \quad + \text{ C}_6\text{H}_5\text{NH}_2 + \text{NH}_3 + 2 \text{ H}_2\text{O}$$

glucosazone

h)

$$\begin{array}{c} \text{CH}=\text{O} \\ \text{HCOH} \\ \text{HOCH} \\ \text{HCOH} \\ \text{HCOH} \\ \text{CH}_2\text{OH} \end{array} \quad + \text{ 2 Cu}^{\text{II}} + \text{4 OH}^- \longrightarrow$$

$$\begin{array}{c} \text{CO}_2\text{H} \\ \text{HCOH} \\ \text{HOCH} \\ \text{HCOH} \\ \text{HCOH} \\ \text{CH}_2\text{OH} \end{array} \quad + \text{ Cu}_2\text{O} + 2 \text{ H}_2\text{O}$$

16-3. *a)* $OH > \overset{\displaystyle H}{\underset{\displaystyle OH}{-\overset{|}{\underset{|}{C}}-}}(CHOH)_3CHO > \overset{\displaystyle H}{\underset{\displaystyle OH}{-\overset{|}{\underset{|}{C}}-H}} > H.$

$$
\begin{array}{c}
CHO \\
| \\
(CHOH)_3 \\
H \diagup \overset{R}{\underset{|}{C}} \diagdown CH_2OH \\
OH
\end{array}
\equiv
\begin{array}{c}
CHO \\
| \\
(CHOH)_3 \\
H \!-\!\!-\!\!-\! OH \\
| \\
CH_2OH
\end{array}
$$

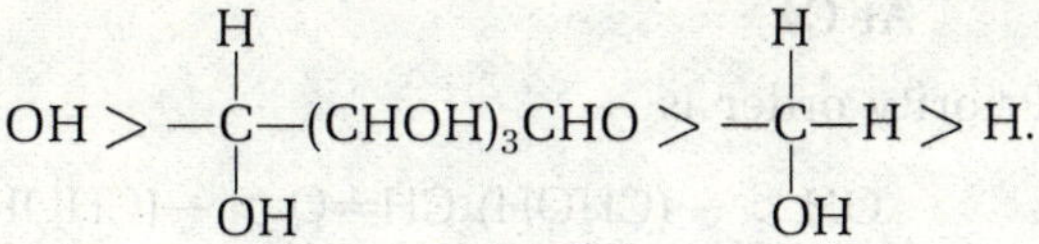

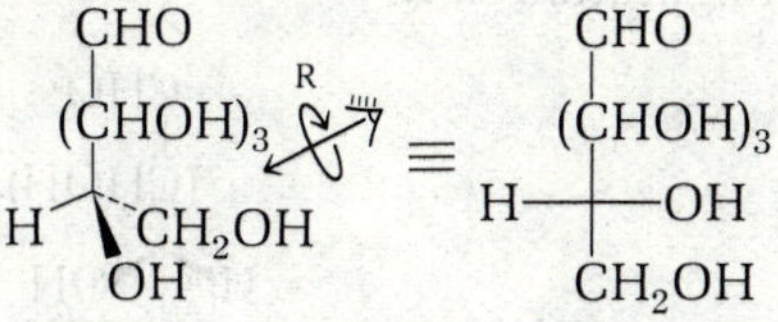

Fischer projection

b) **At C2**

Priority order is

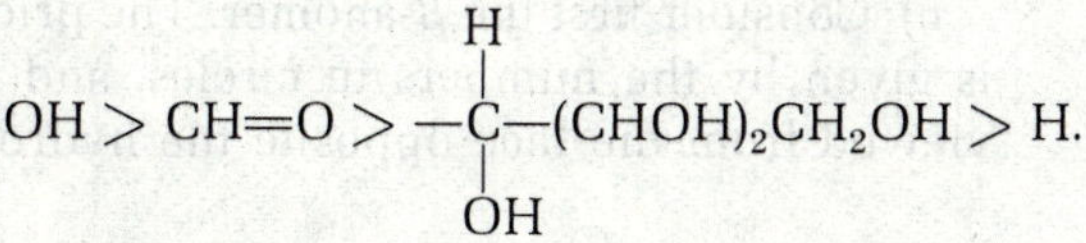

$$OH > CH{=}O > \overset{\displaystyle H}{\underset{\displaystyle OH}{-\overset{|}{\underset{|}{C}}-}}(CHOH)_2CH_2OH > H.$$

Configuration is **R**.

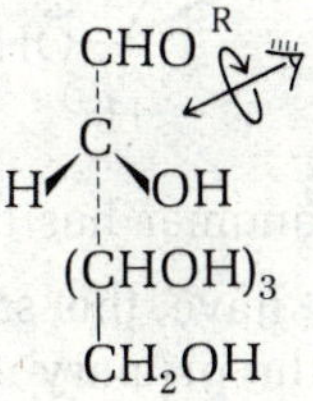

$$
\begin{array}{c}
CHO \quad^{R} \\
\overset{|}{C} \\
H \diagup \quad \diagdown OH \\
(CHOH)_3 \\
| \\
CH_2OH
\end{array}
$$

At C3

Priority order is

$$OH > \overset{\displaystyle H}{\underset{\displaystyle OH}{-\overset{|}{\underset{|}{C}}-}}CH{=}O > \overset{\displaystyle H}{\underset{\displaystyle OH}{-\overset{|}{\underset{|}{C}}-}}(CHOH){-}CH_2OH > H$$

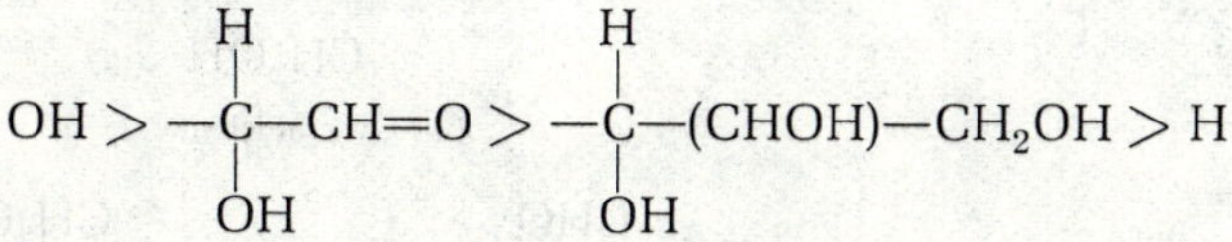

Configuration is **S**

$$
\begin{array}{c}
CHO \\
CHOH \\
HO \diagup \quad \diagdown H \\
(CHOH)_2 \\
| \\
CH_2OH
\end{array}
$$

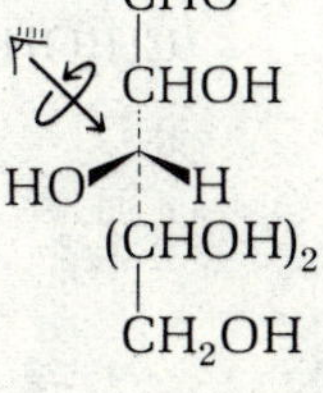

At C4

Priority order is

$$OH > -(CHOH)_2CH{=}O > -(CHOH)CH_2OH > H$$

Configuration is **R**

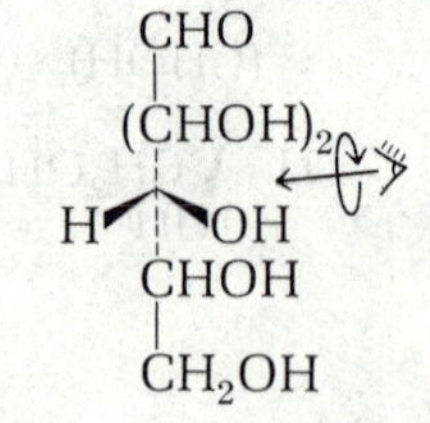

$$
\begin{array}{c}
CHO \\
(CHOH)_2 \\
H \quad OH \\
CHOH \\
CH_2OH
\end{array}
$$

D(+)-Glucose has the **R** configuration at C2, C4 and C5, and the **S** configuration at C3.

c) Consider first the β-anomer. The priority order of the groups is given by the numbers in circles, and the anomeric carbon is viewed from the face opposite the hydrogen:

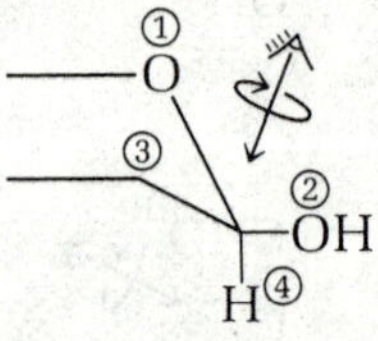

The configuration is **R**.

Since the α-anomer has the opposite configuration, it must be **S**.

16-4. D-Sugars have the same configuration at the carbon atom adjacent to the primary alcohol function as D(+)-glyceraldehyde (which is **R**).

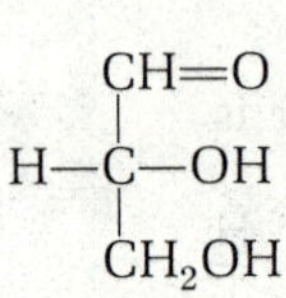

$$
\begin{array}{c}
CH{=}O \\
H-C-OH \\
CH_2OH
\end{array}
$$

D(+)-glyceraldehyde

$$
\begin{array}{c}
CHO \\
(CHOH)_n \\
H-C-OH \\
CH_2OH
\end{array}
\qquad
\begin{array}{c}
CH_2OH \\
C{=}O \\
(CHOH)_n \\
H-C-OH \\
CH_2OH
\end{array}
$$

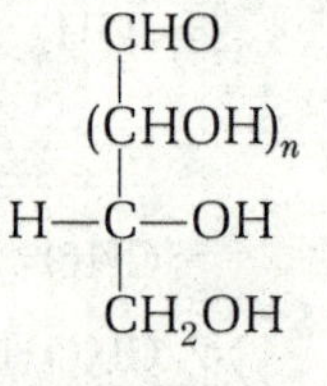

a D-aldose 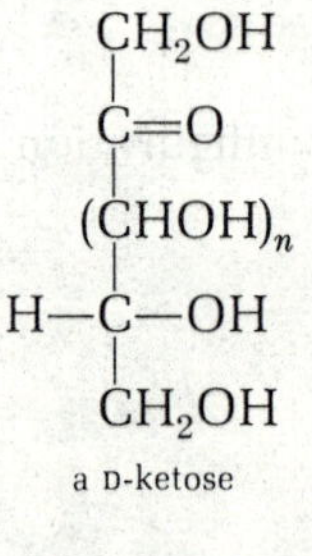   a D-ketose

16-5.

$$\text{(structure, pyranose ring with CHOH, CHOH, CHOH, O, CH, CH}_2\text{OH)} + H^+ \rightleftharpoons \text{(protonated structure with } C-\overset{+}{\underset{\cdot\cdot}{O}}\langle{}^H_H)$$

$$\downarrow -H_2O$$

$$\text{CH—OCH}_3 \text{ structure} \xrightarrow{-H^+} \text{CH—}\overset{+}{\underset{\cdot\cdot}{O}}\langle{}^{CH_3}_{H} \xrightleftharpoons{CH_3\overset{\cdot\cdot}{O}H} \text{CH}^+ \text{ :O: structure} \longleftrightarrow \text{CH, O:}^+ \text{ structure}$$

The first protonation equilibrium can and does occur at any of the oxygens in the molecule. But loss of any other protonated hydroxyl group gives only an ordinary primary or secondary carbonium ion, whereas loss of the protonated hydroxyl group at C1 leads to an oxygen-stabilized carbonium ion, as shown (see also, sec 9.6c).

16-6. Consider starting with the β-anomer:

$$\text{(chair structure with } CH_2OH, HO, HO, OH, OH) \xrightleftharpoons{H^+} \text{(chair structure with } \overset{+}{O}H_2)$$

$$\downarrow -H_2O$$

$$\text{(oxocarbonium ion chair structure)} \longleftrightarrow \text{(oxocarbonium ion chair structure with } \overset{+}{H})$$

This same carbonium ion will be the intermediate, whether one starts with the α- or β-anomer. It can react with a nucleophile (methanol) from either "face" to give both α- and β-products as is shown on the following page.

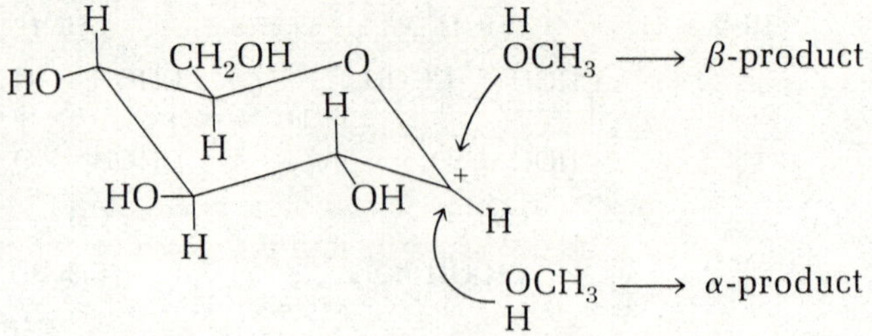

β-product

α-product

16-7. For clarity, only the groups involved in the reaction will be shown.

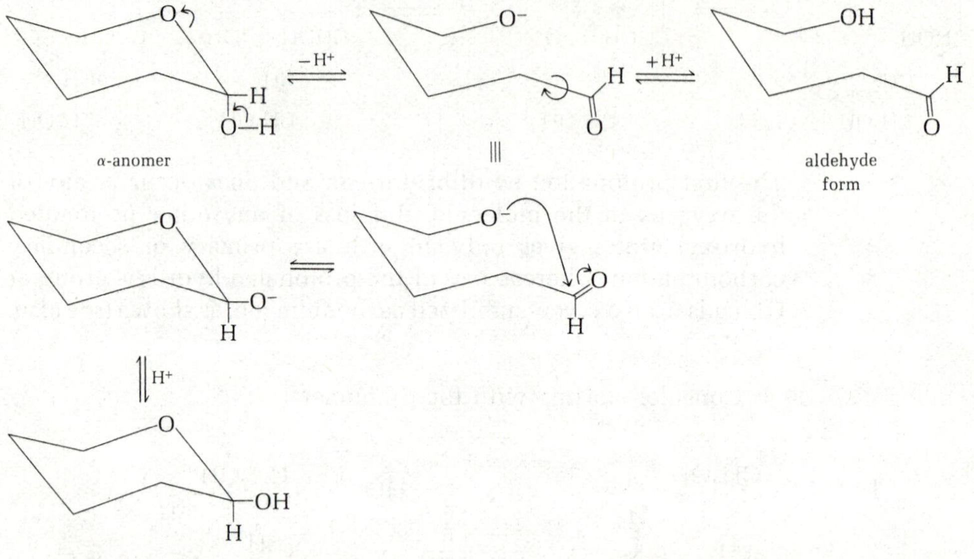

α-anomer

aldehyde form

β-anomer

The first step, involving proton loss from the hemiacetal hydroxyl group, is not possible when the proton is replaced by an alkyl group.

16-8. The structure is

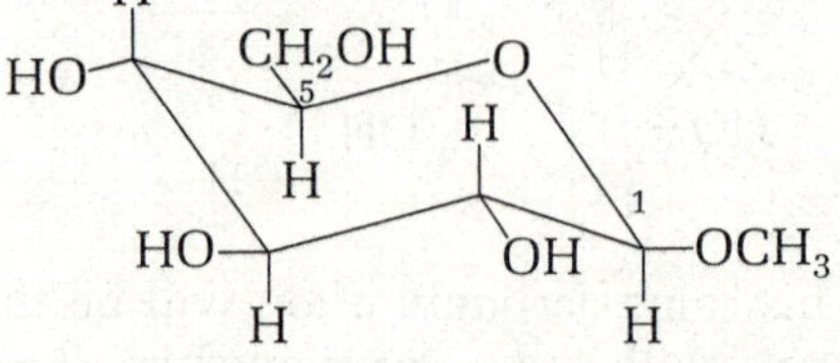

Methyl refers to the name of the alkyl group attached to the oxygen at C1.

β refers to the configuration of the methoxyl group at C1 ("up," as shown).

D refers to the configuration at C5, which is the same as that of D-glyceraldehyde.

(+) means that the compound is dextrorotatory.

Gluco refers to glucose, and indicates the configurations at C2, C3 and C4.

Pyranos indicates that the oxygen-containing heterocyclic ring is 6-membered.

Ide; this suffix indicates that the compound is an acetal (a glycoside).

16-9. The structures are mirror images of those shown in Table 16-2.

$$
\begin{array}{ccc}
& \text{CH=O} & \\
\text{HO} & \!\!-\!\!| & \text{H} \\
\text{H} & |\!\!-\!\! & \text{OH} \\
\text{HO} & \!\!-\!\!| & \text{H} \\
\text{HO} & \!\!-\!\!| & \text{H} \\
& \text{CH}_2\text{OH} &
\end{array}
\qquad
\begin{array}{ccc}
& \text{CH=O} & \\
\text{HO} & \!\!-\!\!| & \text{H} \\
\text{HO} & \!\!-\!\!| & \text{H} \\
\text{HO} & \!\!-\!\!| & \text{H} \\
& \text{CH}_2\text{OH} &
\end{array}
$$

L(−)-glucose   L(+)-ribose

16-10. The first steps are shown in Table 16-2. Oxidation of D-erythrose will give meso-tartaric acid,

$$
\begin{array}{ccc}
& \text{CHO} & \\
\text{H} & |\!\!-\!\! & \text{OH} \\
\text{H} & |\!\!-\!\! & \text{OH} \\
& \text{CH}_2\text{OH} &
\end{array}
\quad\longrightarrow\quad
\begin{array}{ccc}
& \text{CO}_2\text{H} & \\
\text{H} & |\!\!-\!\! & \text{OH} \\
\text{H} & |\!\!-\!\! & \text{OH} \\
& \text{CO}_2\text{H} &
\end{array}
$$

D-erythrose   meso-tartaric acid
(optically inactive)

whereas analogous oxidation of D-threose will give an optically active tartaric acid:

$$
\begin{array}{ccc}
& \text{CHO} & \\
\text{HO} & \!\!-\!\!| & \text{H} \\
\text{H} & |\!\!-\!\! & \text{OH} \\
& \text{CH}_2\text{OH} &
\end{array}
\quad\longrightarrow\quad
\begin{array}{ccc}
& \text{CO}_2\text{H} & \\
\text{HO} & \!\!-\!\!| & \text{H} \\
\text{H} & |\!\!-\!\! & \text{OH} \\
& \text{CO}_2\text{H} &
\end{array}
$$

D-threose   optically active
(S,S) tartaric acid

In this way one can readily assign structures to the two tetroses.

16-11. *a)*

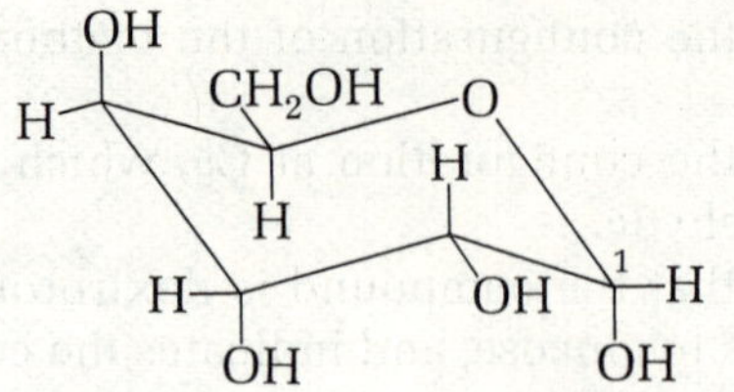

Compare the configurations at each carbon atom with those of glucose.

b) 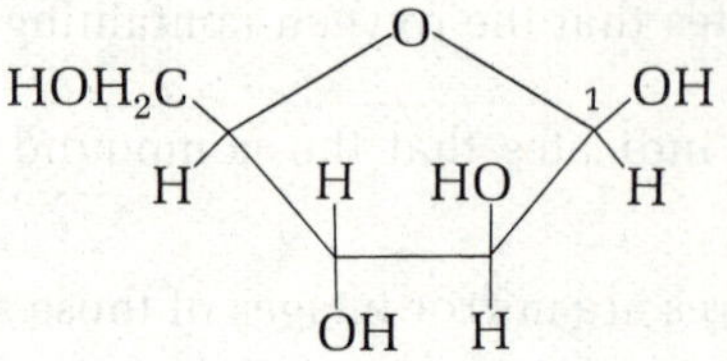 HOH$_2$C

16-12. Use Table 16-2 as a guide.

Since D-ribose and D-arabinose have the same configurations at C3 and C4 they will give the same osazone, with the structure

$$
\begin{array}{l}
CH\!=\!NNHC_6H_5 \\
| \\
C\!=\!NNHC_6H_5 \\
| \\
HCOH \\
| \\
HCOH \\
| \\
CH_2OH
\end{array}
$$

Similarly D-xylose and D-lyxose will give the identical osazone, with the structure

$$
\begin{array}{l}
CH\!=\!NNHC_6H_5 \\
| \\
C\!=\!NNHC_6H_5 \\
| \\
HOCH \\
| \\
HCOH \\
| \\
CH_2OH
\end{array}
$$

Using analogous reasoning, the eight D-hexoses in Table 16-2 will give four osazones, each of the following pairs giving the same osazone.

D-allose	D-glucose	D-gulose	D-galactose
D-altrose	D-mannose	D-idose	D-talose

16-13. The mechanism is identical for both glycosides, and will be shown for only one.

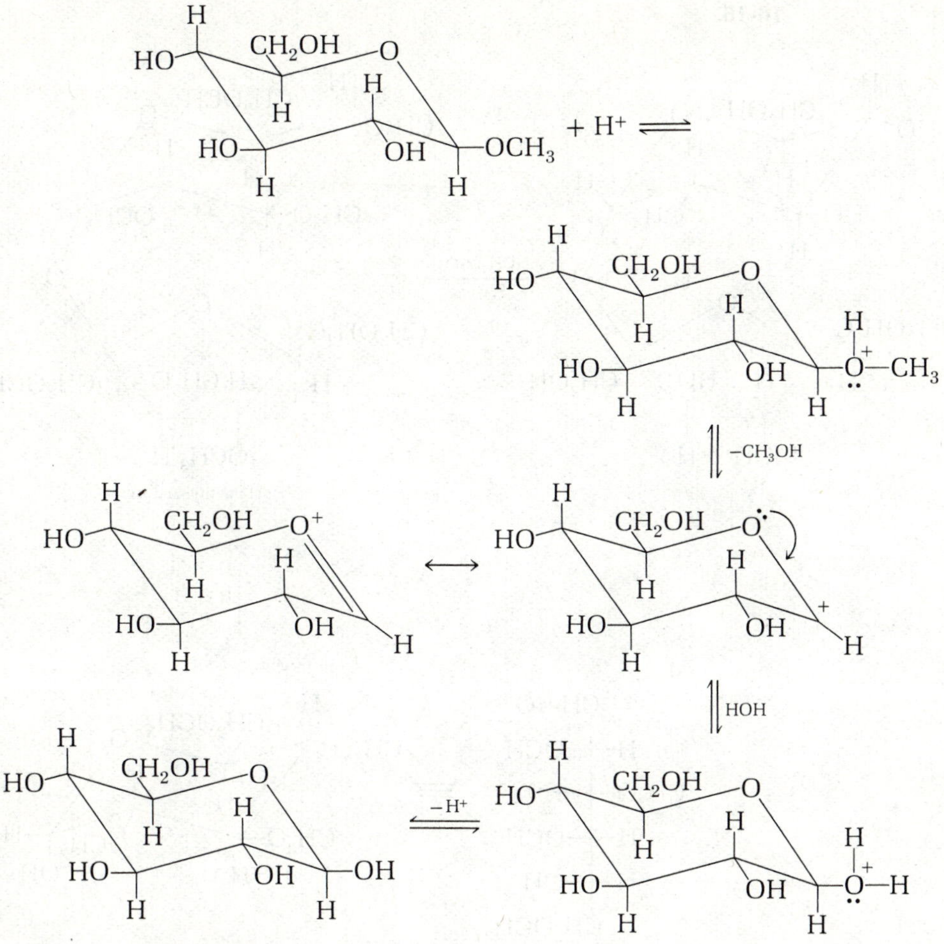

A mixture of the
α and β forms
of glucose is obtained.

There is no acidic proton at C1 in a glycoside (i.e., the compound is an acetal, not a hemiacetal). Consequently hydrolysis cannot be catalyzed by a base. But the unshared electron pairs on the alkoxyl group at C1 furnish a site for acid catalysis, as shown above.

16-14. At equilibrium the specific rotation of D-glucose is $+52°$, whereas the pure α and β forms have specific rotations of $+113°$ and $+19°$ respectively. The % β form at equilibrium will therefore be

$$\frac{113-52}{113-19} \times 100 = \frac{61}{94} \times 100 = 64.8\%.$$

16-15.

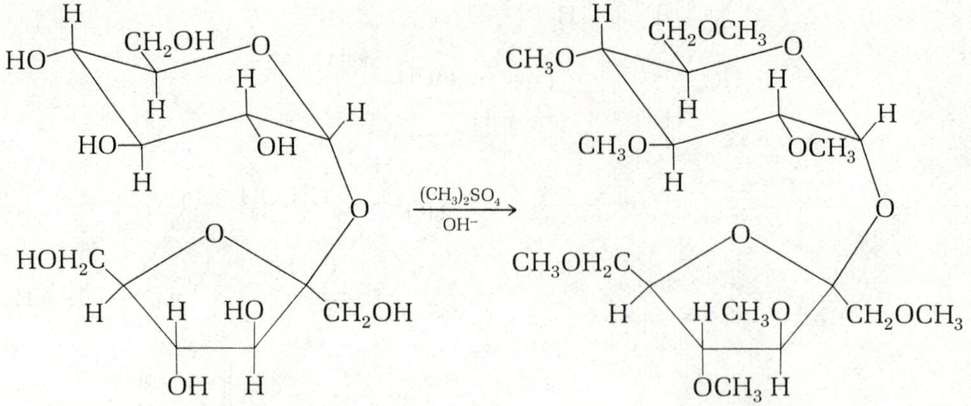

octamethylsucrose

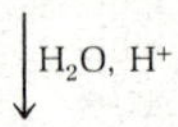

$\downarrow$ H$_2$O, H$^+$

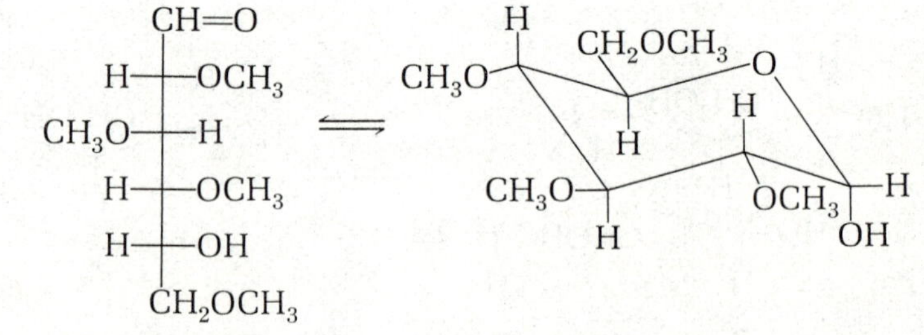

2,3,4,6-tetramethylglucose
(Fischer projection)

2,3,4,6-tetramethylglucose
(cyclic structure)

+

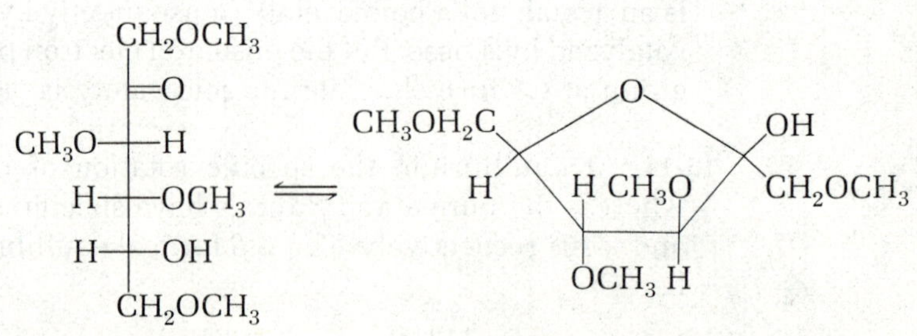

1,3,4,6-tetramethylfructose
(Fischer projection)

1,3,4,6-tetramethylfructose
(furanose structure)

16-16.

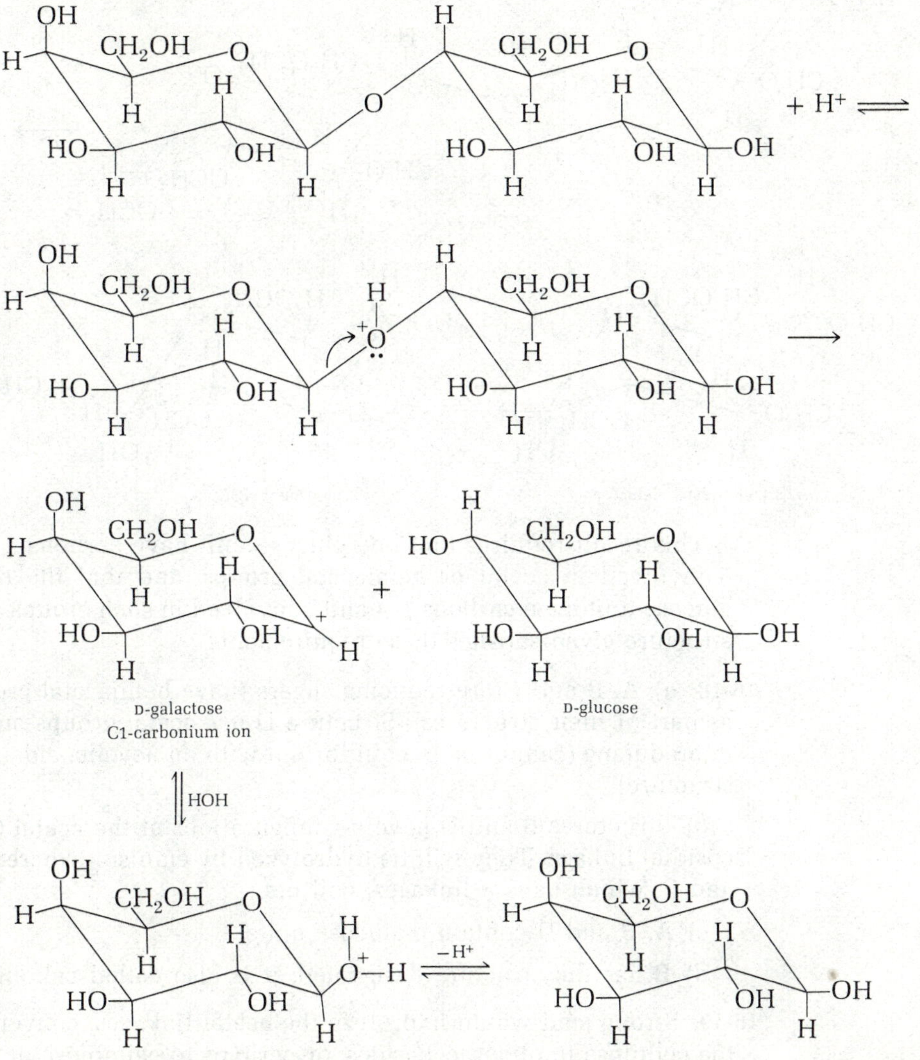

D-galactose
C1-carbonium ion

D-glucose

HOH

D-galactose
(β-form; the α-form is
also obtained).

16-17. There are eight hydroxyl groups in maltose; methylation with
dimethyl sulfate will therefore give octamethylmaltose. The compound contains *two* acetal-type carbon atoms, marked with asterisks. Acidic hydrolysis occurs at both these sites as shown on
the following page.

2,3,4,6-tetramethylglucose   2,3,6-trimethylglucose

This result requires that one glucose unit have carbons 1 and 5 involved in acetal or hemiacetal groups, and that the other glucose unit have carbons 1, 4 and 5 involved in such groups. The structure given satisfies these requirements.

16-18. *a*) A, B and C are reducing sugars (have hemiacetal groups as part of their structures). Structure D has acetal groups and is nonreducing (cannot be in equilibrium with an acyclic, aldehyde structure).

b) Structures B and D have β-configurations at the acetal (glycosidic) linkage. They will be hydrolyzed by emulsin, whereas A and C, which have α linkages, will not.

c) A, B and D contain D-glucose units.

d) B has the structure designated; it is also called cellobiose.

16-19. Strong acid would hydrolyze the acetal linkages, converting the cellulose to oligosaccharides, or perhaps to cellobiose or glucose.

16-20. *a*) **Eq 16-18**

The reaction involves nucleophilic attack on the carbon-sulfur double bond, entirely analogous to reactions of carbonyl compounds with nucleophiles.

Eq 16-19

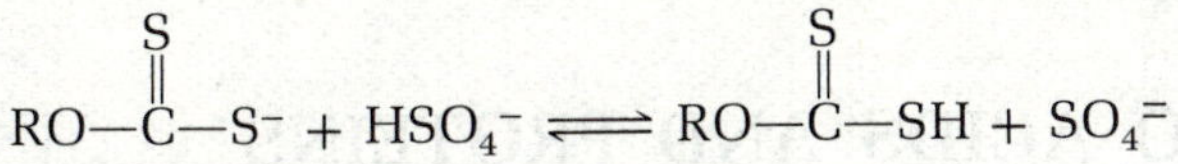

$$RO-\overset{\overset{S}{\|}}{C}-S^- + HSO_4^- \rightleftharpoons RO-\overset{\overset{S}{\|}}{C}-SH + SO_4^=$$

Protonation may even proceed further:

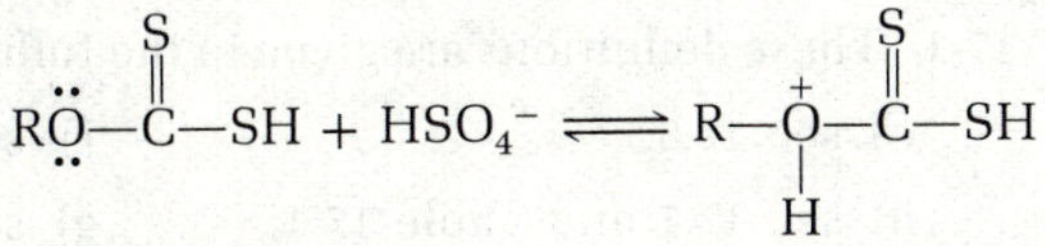

$$R\overset{..}{\underset{..}{O}}-\overset{\overset{S}{\|}}{C}-SH + HSO_4^- \rightleftharpoons R-\overset{+}{\underset{\underset{H}{|}}{O}}-\overset{\overset{S}{\|}}{C}-SH$$

A 1,2-elimination then can occur:

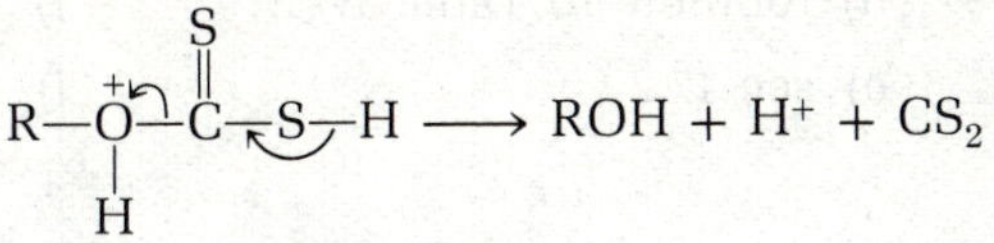

$$R-\overset{+}{\underset{\underset{H}{|}}{O}}-\overset{\overset{S}{\|}}{C}-S-H \longrightarrow ROH + H^+ + CS_2$$

b)

$$+ CS_2 + NaOH \longrightarrow$$

$$\xrightarrow{H^+}$$

$$+ CS_2$$

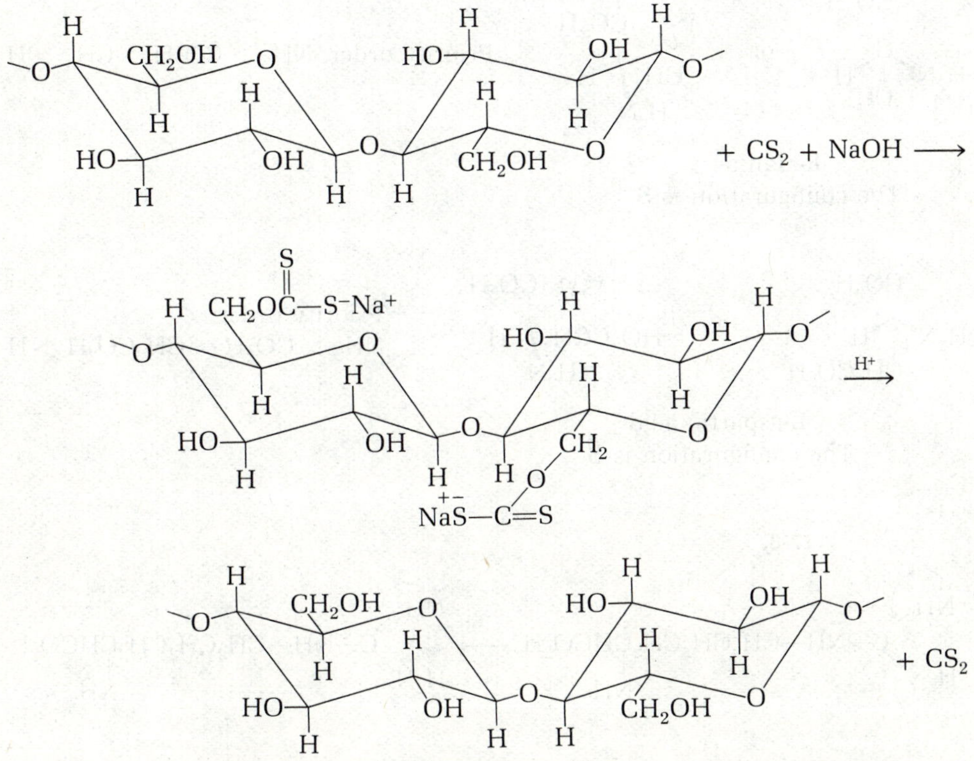

The reaction can occur at any of the three hydroxyl groups in each cellulose unit. For simplicity, it is shown above occurring only at the primary alcohol group of each unit.

CHAPTER SEVENTEEN

AMINO ACIDS AND PROTEINS

17-1. These definitions are given in the following sections of the text:

a) sec 17.6	f) sec 17.2
b) sec 17.1 and Table 17-1.	g) sec 17.1, also sec 16.2a
c) sec 17.1 and Table 17-1.	h) sec 17.6
d) footnote to Table 17-1.	i) sec 17.6
e) sec 17.2	j) sec 17.6

17-2.

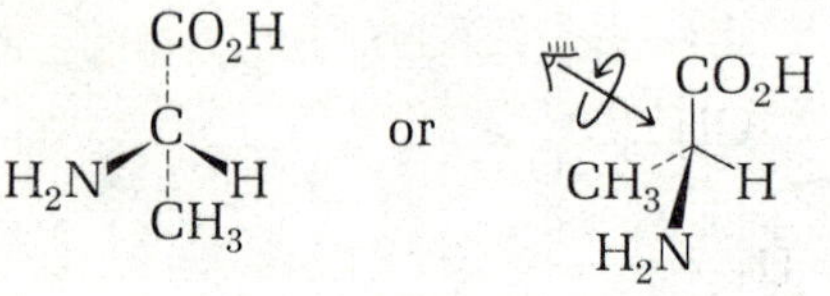

Priority order: $NH_2 > CO_2H > CH_3 > H$

L-alanine
The configuration is **S**.

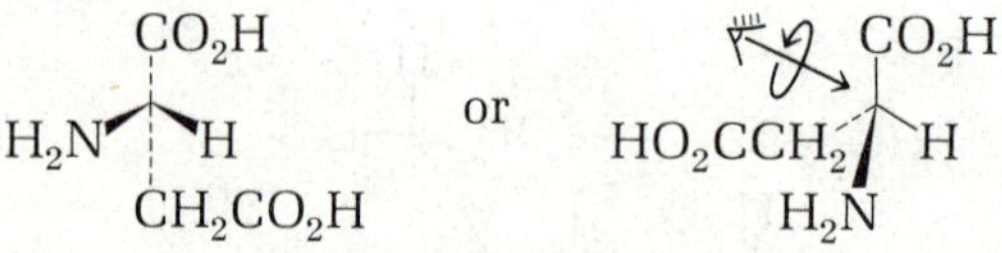

Priority order:
$NH_2 > CO_2H > CH_2CO_2H > H$

L-aspartic acid
The configuration is **S**.

17-3.

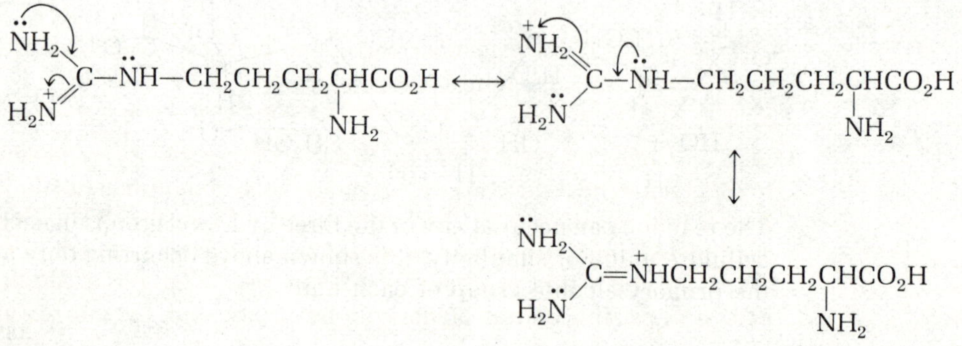

The positive charge is delocalized over three nitrogen atoms, whereas protonation of an ordinary amine RNH_2 places the positive charge on a single nitrogen atom, as in $R\overset{+}{N}H_3$. Thus the energy difference between a guanidine and its protonated form is less than the energy difference between a simple amine and its protonated form.

17-4. We will write alanine in the dipolar form:

a) $CH_3\underset{\overset{|}{NH_3^+}}{CH}CO_2^- + H^+Cl^- \longrightarrow CH_3\underset{\overset{|}{NH_3^+}\ Cl^-}{CH}CO_2H$

> The ammonium ion, being positive, migrates toward the cathode.

b) $CH_3\underset{\overset{|}{NH_3^+}}{CH}CO_2^- + Na^+OH^- \longrightarrow CH_3\underset{\overset{|}{NH_2}}{CH}CO_2^-Na^+ + H_2O$

> The carboxylate ion, being negative, migrates toward the anode.

17-5. Lysine exists as the negative ion in strongly alkaline solutions. Acidification results in stepwise addition of three protons:

$$H_2NCH_2CH_2CH_2\underset{\overset{|}{NH_2}}{CH}CO_2^- + H^+ \rightleftharpoons$$

$$H_3\overset{+}{N}CH_2CH_2CH_2\underset{\overset{|}{NH_2}}{CH}CO_2^- \text{ and } H_2NCH_2CH_2CH_2\underset{\overset{|}{NH_3^+}}{CH}CO_2^-$$

In its dipolar form, lysine is protonated mainly on the terminal amino group. Further acidification gives $H_3\overset{+}{N}CH_2CH_2CH_2\underset{\overset{|}{NH_3^+}}{CH}CO_2^-$

and eventually $H_3\overset{+}{N}CH_2CH_2CH_2\underset{\overset{|}{NH_3^+}}{CH}CO_2H.$

The dipolar stage is reached when only one proton is added, appreciably before the solution is neutral (*pH* 7). At neutrality, lysine exists mainly as the monopositive ion.

17-6. The isoelectric points of the three amino acids are expected to be: glycine (near neutral), arginine (basic), glutamic acid (acidic). The actual values are *pH* 5.97, 10.76, and 3.22, respectively. See section 17.3 for a discussion of amino acid separation which will help you solve this problem. Since the resin contains sulfonic acid groups, the amino acids will be protonated by the resin.

Arginine, being most basic, will be most easily protonated and most firmly bound to the resin. Glycine will be next; and glutamic acid, with two carboxyl groups, will be least readily protonated and therefore least firmly bound to the resin. The order of elution will then be the inverse of the above; glutamic acid will come off the column first, glycine next, and arginine last.

17-7. Write equations 17-4 and 17-5, with R= ![benzyl group structure]—CH₂—.

17-8. Ninhydrin is in equilibrium with the triketo structure:

Nucleophilic attack of the amino nitrogen occurs next (compare with eq 9-28).

The mechanism of the next step, decarboxylation, is shown in equation 17-5. Hydrolysis occurs as follows:

This structure is a nitrogen analog of a hemiacetal.

The final step is analogous to the first.

blue dye

17-9. *a)* $CH_3CH_2CO_2H \xrightarrow[\substack{PCl_3 \\ eq\ 14\text{-}22}]{Cl_2} CH_3\underset{\underset{Cl}{|}}{C}HCO_2H \xrightarrow[\substack{excess \\ eq\ 17\text{-}6}]{NH_3} CH_3\underset{\underset{NH_2}{|}}{C}HCO_2H$

b) $(CH_3)_2CHCH{=}O \xrightarrow[NH_3]{HCN} (CH_3)_2CH\overset{\overset{NH_2}{|}}{C}HCN \xrightarrow[H^+]{H_2O}$

$(CH_3)_2CH\overset{\overset{NH_2}{|}}{C}HCO_2H$ (eq 17-7)

c) $CH_3CO_2H \xrightarrow[\substack{PCl_3 \\ eq\ 14\text{-}22}]{Cl_2} \underset{\underset{Cl}{|}}{C}H_2CO_2H$, then follow equation 17-6.

d) Many routes are possible. One must add two carbons to the chain. One path is:

$$(CH_3)_2CHCH_2Cl \xrightarrow[\substack{(2)\ CH_2O}]{\substack{(1)\ Mg,\ ether}} (CH_3)_2CHCH_2CH_2OH$$
$$\text{(eq 9-35)}$$

$$\downarrow \begin{array}{l} Cr_2O_7^= \quad \text{(eq 9-5)} \\ H^+ \end{array}$$

$$(CH_3)_2CHCH_2\underset{\underset{NH_2}{|}}{C}H-CO_2H \xleftarrow[\substack{(2)\ H_2O,\ H^+}]{\substack{(1)\ HCN,\ NH_3}} (CH_3)_2CHCH_2CHO$$
$$\text{(eq 17-7)}$$

17-10. *a*) $CH_3\underset{\underset{NH_2}{|}}{C}HCO_2H + HONO \xrightarrow[\text{(eq 17-8)}]{} CH_3\underset{\underset{OH}{|}}{C}HCO_2H + H_2O + N_2$

lactic acid

b) $CH_3\underset{\underset{NH_2}{|}}{C}HCO_2H +$ $\xrightarrow[\text{(eq 12-26)}]{}$

$$CH_3\underset{\underset{\underset{O}{\overset{||}{C}}}{\underset{|}{NH}}}{C}HCO_2H + HCl$$

c) $2\ CH_3\underset{\underset{NH_2}{|}}{C}HCO_2H \xrightarrow[\text{(eq 17-11)}]{\text{heat}}$

d) $CH_3\underset{\underset{NH_2}{|}}{C}HCO_2H + (CH_3CO)_2O \xrightarrow[\text{(eq 12-24)}]{}$

$$CH_3\underset{\underset{\underset{O}{\overset{||}{C}}}{\underset{|}{NH}}CH_3}{C}HCO_2H + CH_3CO_2H$$

e) $CH_3\underset{\underset{NH_2}{|}}{C}HCO_2H + CH_3CH_2OH \xrightarrow[\text{(eq 17-9)}]{H^+}$

$$CH_3\underset{\underset{NH_2}{|}}{C}HCO_2CH_2CH_3 + H_2O$$

17-11. In each case one can usually begin by writing down the backbone of the peptide chain:

$$\overset{+}{H_3N}-CH-\overset{\overset{\displaystyle O}{\|}}{C}-NH-CH-\overset{\overset{\displaystyle O}{\|}}{C}-NH-CH-\overset{\overset{\displaystyle O}{\|}}{C}- \text{ etc.}$$

Then fill in the appropriate R groups.

a) $\overset{+}{H_3N}CH_2\overset{\overset{\displaystyle O}{\|}}{C}\ NHCH\overset{\overset{\displaystyle O}{\|}}{C}\ NHCH_2CO_2{}^-$

$\qquad\qquad\qquad CH_3$

b) $\quad H_2N-CH-\overset{\overset{\displaystyle O}{\|}}{C}-NH-CH-CO_2{}^-$

$\qquad CH{=}C-CH_2 \qquad\qquad CH(CH_3)_2$

$\quad HN^+ \qquad NH$

$\qquad\diagdown\!\!\diagup$

$\qquad\quad CH$

c) $\overset{+}{H_3N}-CH-\overset{\overset{\displaystyle O}{\|}}{C}-NH-CH-\overset{\overset{\displaystyle O}{\|}}{C}-NH-CH-CO_2{}^-$

$CH_3-CH-CH_2CH_3 \quad CH_2\,CH\,(CH_3)_2\,CH_2OH$

d) $\overset{+}{H_3N}-CH-\overset{\overset{\displaystyle O}{\|}}{C}-NH-CH-CO_2{}^-$

$\qquad CH_2CO_2H \qquad CH_2$

(indole ring with N–H)

17-12. There are six possibilities altogether:

Gly—Ala—Ser
Gly—Ser—Ala
Ala—Gly—Ser
Ala—Ser—Gly
Ser—Gly—Ala
Ser—Ala—Gly

The full structure of the last of these is:

$$\overset{+}{H_3N}-CH-\overset{\overset{\displaystyle O}{\|}}{C}-NH-CH-\overset{\overset{\displaystyle O}{\|}}{C}-NH-CH-CO_2{}^-$$

$\qquad CH_2OH \qquad\qquad CH_3 \qquad\qquad H$

17-13. The structure is identical with that of oxytocin, except for the replacement of isoleucine by phenylalanine and leucine by arginine (sec 17.6). The structures are:

$$Cy-S \qquad \begin{array}{c} Tyr \\ \end{array} \qquad Phe$$

```
                Tyr
        Cy—S          Phe
        Cy—S         Glu—NH₂
    Pro         Asp—NH₂
          Arg—Gly—NH₂
```

17-14.

17-15. The overlapping sequences are easily discerned, as is shown on the facing page. The peptide must be Ala—Gly—Val—Tyr—Cys—Phe—Leu—Try, an octapeptide. The N-terminal acid is alanine, the C-terminal is tryptophan and the name is: alanylglycylvalyltyrosylcystylphenylalanylleucyltryptophan.

> Ala—Gly
> Gly—Val
> Gly—Val—Tyr
> Val—Tyr—Cys
> Tyr—Cys—Phe
> Cys—Phe—Leu
> Phe—Leu—Try

17-16. For a model, see section 8.5, especially equation 8-30.

2,4-dinitrophenylglycine

17-17. a) Trypsin hydrolyzes proteins at the carbonyl end of lysine and arginine residues (sec 17.7). The expected fragments are:

Ala—Lys   Glu—Leu—Phe—Lys   Tyr—Val—Arg   and   Gly

b) Chymotrypsin hydrolyzes proteins at the carbonyl end of tyrosine and phenylalanine residues (sec 17.7). The expected fragments are:

Ala—Lys—Glu—Leu—Phe   Lys—Tyr   Val—Arg—Gly

c) Carboxypeptidase hydrolyzes peptide bonds adjacent to free carboxyl groups (sec 17.7). The expected fragments are:

Ala—Lys—Glu—Leu—Phe—Lys—Tyr—Val—Arg and Gly.

d) Aminopeptidase hydrolyzes peptide bonds adjacent to free amino groups (sec 17.7). The expected fragments are:

Ala and Lys—Glu—Leu—Phe—Lys—Tyr—Val—Arg—Gly.

17-18. The principal factor in secondary protein structure (sec 17.8) is hydrogen bonding. Factors leading to tertiary structure are disulfide linkages, electrostatic interactions between positive ($-\overset{+}{N}H_3$) and negative ($-CO_2^-$) groups, hydrogen bonds between functions on the R-groups of amino acid side-chains, and hydrophobic attractions between nonpolar alkyl side-chains. Factors

193

leading to quaternary structures are less well understood, but include many of the above.

17-19.

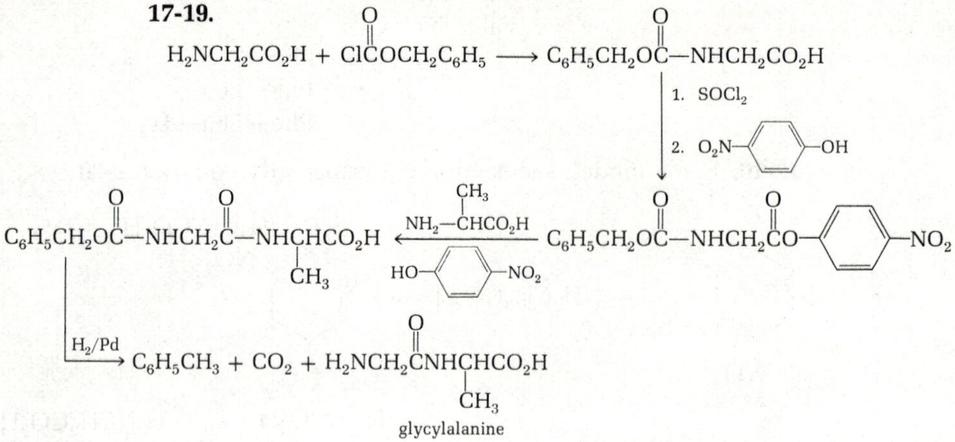

$$H_2NCH_2CO_2H + ClCOCH_2C_6H_5 \longrightarrow C_6H_5CH_2OC-NHCH_2CO_2H$$

1. SOCl₂

2. O₂N—⟨ ⟩—OH

$$C_6H_5CH_2OC-NHCH_2C-NHCHCO_2H \xleftarrow{\substack{NH_2-CHCO_2H \\ \\ HO-⟨ ⟩-NO_2}} C_6H_5CH_2OC-NHCH_2CO-⟨ ⟩-NO_2$$

with CH₃ substituents

$\downarrow$ H₂/Pd

$$\longrightarrow C_6H_5CH_3 + CO_2 + H_2NCH_2CNHCHCO_2H$$

CH₃

glycylalanine

17-20. The first result tells us that the N-terminal amino acid is methionine; the structure at this point is

Met(2 Met, Ser, Gly)

From dipeptide D we learn that one sequence is

Ser—Met

From dipeptide C we learn that one sequence is

Met—Met

From tripeptide B we learn that one sequence is

Met(Met, Ser)

and in view of the result from dipeptide D, tripeptide B must be

Met—Ser—Met

Since two methionines must be adjacent (dipeptide C), the possibilities are

Met—Met—Ser—Met—Gly
and Met—Ser—Met—Met—Gly

Tripeptide A allows a decision, since 2 Met's and one Gly must be joined together.

Therefore the correct structure is

Met—Ser—Met—Met—Gly

CHAPTER EIGHTEEN

NATURAL PRODUCTS

18-1. The definitions, and examples, are given in the following sections of the text:

- a) sec 18.1
- b) sec 18.2
- c) sec 18.2a
- d) sec 18.5a
- e) sec 18.5b
- f) sec 18.5c
- g) sec 18.2b, 17.6
- h) sec 18.6

18-2. We will use dashed lines to divide the molecules into isoprene units.

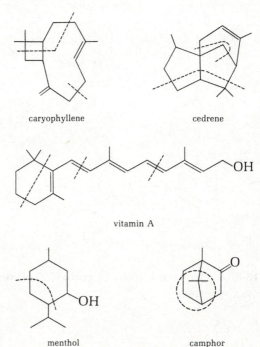

caryophyllene

cedrene

vitamin A

menthol

camphor

18-3. Use the second formula shown in section 18.2 as a guide.

- a) equatorial
- b) equatorial
- c) equatorial
- d) equatorial

The pattern is that equatorial bonds are β at C1, C3, and C7 (the odd-numbered carbons) and α at C2, C4, and C6 (the even-numbered carbons).

18-4. Use the formulas in section 18.2a as guides.

Cholesterol: C3, C8, C9, C10, C13, C14, C17, C20 are chiral. Since all asymmetric carbons are different, there are in principle $2^8 = 256$ possible stereoisomers.

Ergosterol has chiral centers at C3, C9, C10, C13, C14, C17, C20, and C24. Furthermore, the side-chain double bond could be *cis* or *trans*. Therefore there are 2^9 or 512 possible isomers.

18-5. *a*)

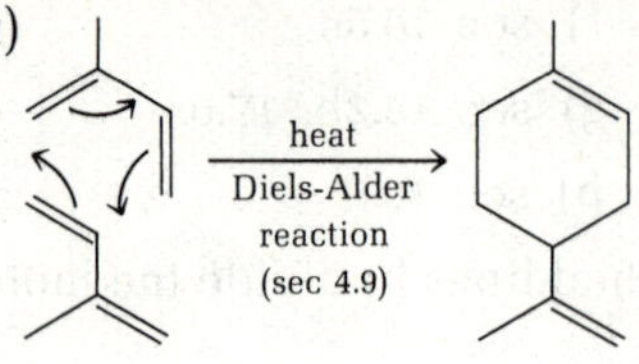

2 isoprenes

b)

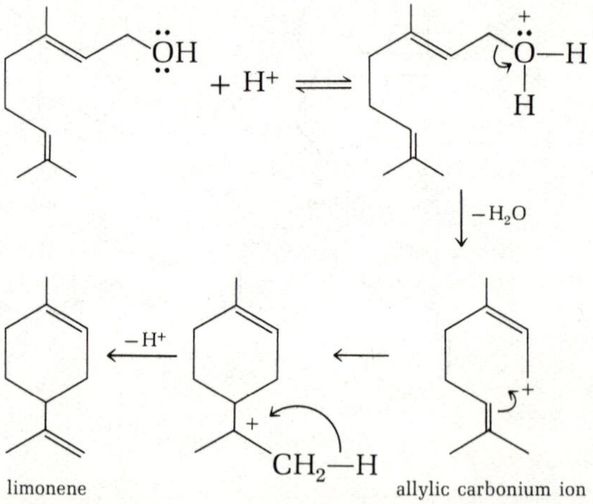

limonene

allylic carbonium ion

18-6. Use equation 18-6 as a guide. Labels will be indicated here with an asterisk.

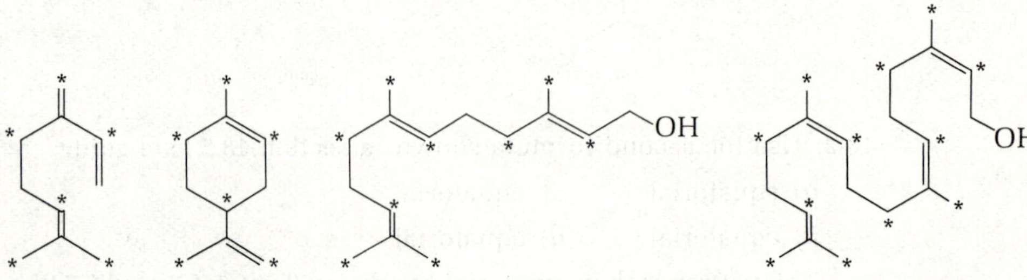

myrcene limonene farnesol geranylgeraniol (eq 18-7)

18-7. The formula for farnesyl pyrophosphate is given in equation 18-7. Begin by loss of a pyrophosphate anion to form the allylic farnesyl carbonium ion.

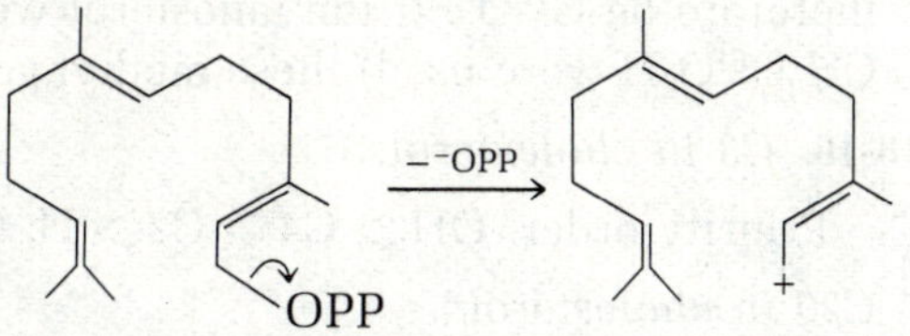

This ion may then give either of the two clove terpenes:

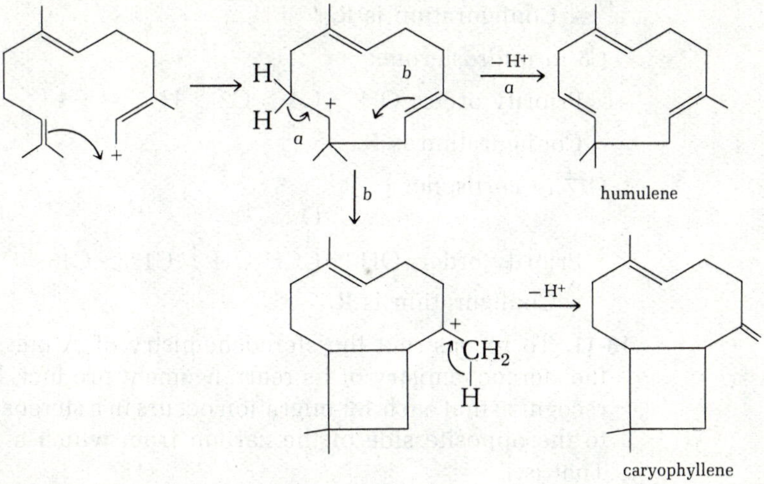

humulene

caryophyllene

18-8. The squalene precursor is labeled as shown in equation 18-8.

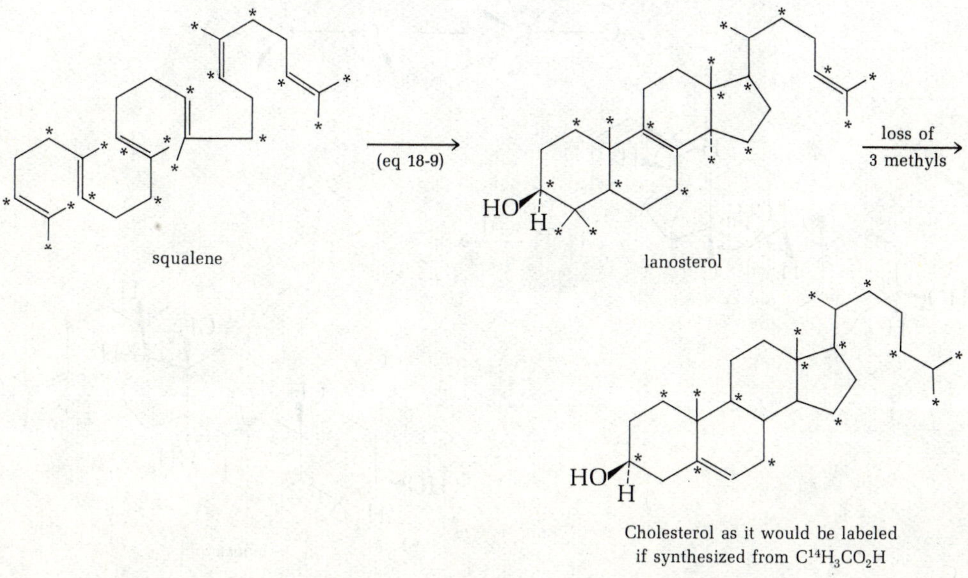

squalene   (eq 18-9)   lanosterol $\xrightarrow{\text{loss of}}$ 3 methyls

Cholesterol as it would be labeled
if synthesized from $C^{14}H_3CO_2H$

18-9. See the final step in the previous answer. The three methyl groups (two at C4 and one at C14) which are lost in the conversion of lanosterol to cholesterol are all labeled. The product would therefore be $C^{14}O_2$. If the lanosterol were counter-labeled (i.e., if $CH_3C^{14}O_2H$ were used) these methyl groups would be unlabeled.

18-10. *C3 in cholesterol:*

Priority order: OH > C4 > C2 > H; Configuration is **S.**

C20 in cholesterol:

Priority order: C17 > C22 > C21 > H

Configuration is **R.**

C3 in androsterone:

Priority order: OH > C4 > C2 > H

Configuration is **R.**

C17 in cortisone:

$$\overset{\displaystyle O}{\overset{\displaystyle \|}{\text{Priority order: OH} > \text{CCH}_2\text{OH} > \text{C13} > \text{C16}}}$$

Configuration is **R.**

18-11. To reconstruct the stereochemistry of A one must examine the stereochemistry of its rearrangement product, lanosterol, and recognize that each 1,2-migration occurs in a stereospecific manner to the opposite side of the carbon from which a group departs. That is:

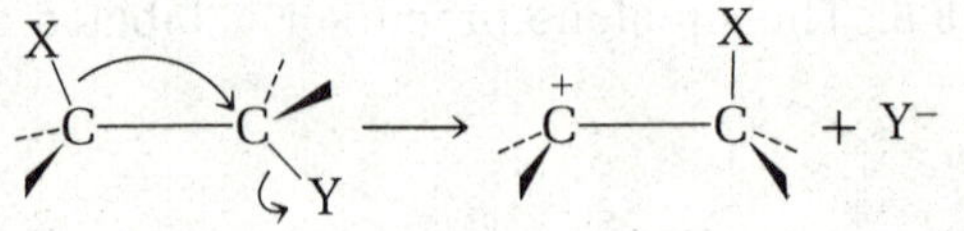

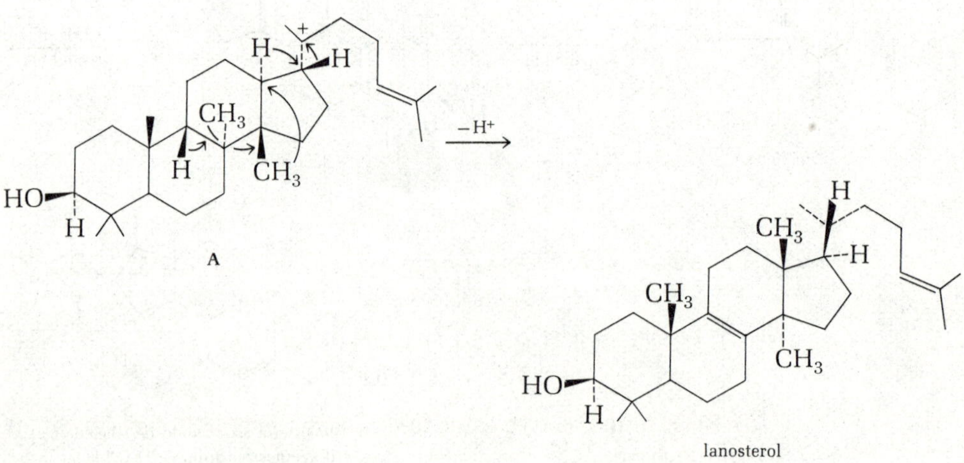

lanosterol

18-12. *a)* Cocaine is a diester; both ester groups would be saponified (eq 10-21).

$$\xrightarrow{\text{2 NaOH}}$$

$$+ \ CH_3OH +$$

b) The ester group in atropine is saponified by dilute base.

$$\xrightarrow{\text{NaOH}}$$

c) Novocaine is an ester; this group will be saponified on heating with dilute base.

$$\xrightarrow{\text{NaOH}}$$

d) Reserpine has two ester functions which might be saponified by dilute base.

$$\xrightarrow[\text{OH}^-]{\text{dil.}}$$

$$+ \quad CH_3CO_2^- + CH_3O- \qquad -CO_2^-$$

18-13. The priority order of groups at the asymmetric carbon atom is:

$$OH > \text{(catechol)} -OH > CH_2NHCH_3 > H$$

The **R** isomer is:

18-14. The equation is (sec 18.5c; also see eq 17-22)

| pyridoxamine | pyruvic acid | pyridoxal | alanine |

A possible mechanism for the transamination reaction is:

$$\text{tautomerism (sec 9.7)}$$

$$\xleftarrow{-H_2O}$$

$$\xrightarrow[\substack{\text{nucleophilic} \\ \text{attack on the} \\ \text{C=N bond}}]{+H_2O}$$

$$\xrightleftharpoons{}$$

This compound is a nitrogen analog of a hemiacetal, and easily reverts to the aldehyde and the amine.

18-15. It seems most likely that the four- and five-membered rings would be *cis*-fused:

In this case, the carboxyl and amide groups can occupy any of four possible relationships (though some would be very crowded sterically). Two of these four possibilities are

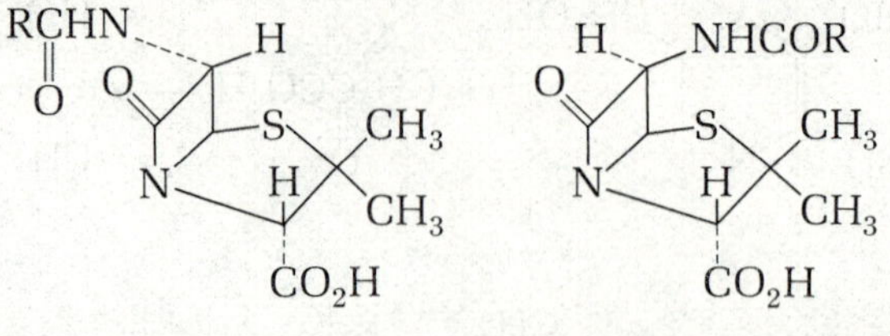

The reader can readily construct the other two. Each of these four has a mirror image which is nonsuperimposable on itself, since each structure is devoid of symmetry. Consequently eight stereoisomers are possible.

18-16. The necessary information is given in section 18.6a.

a) HOCH$_2$

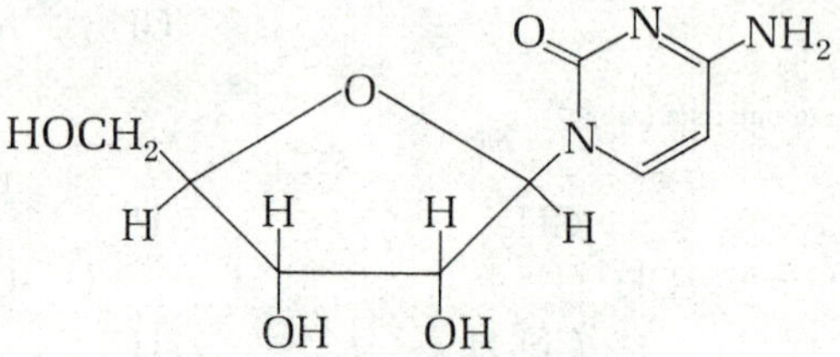

cytidine

b) HOCH$_2$

uridine

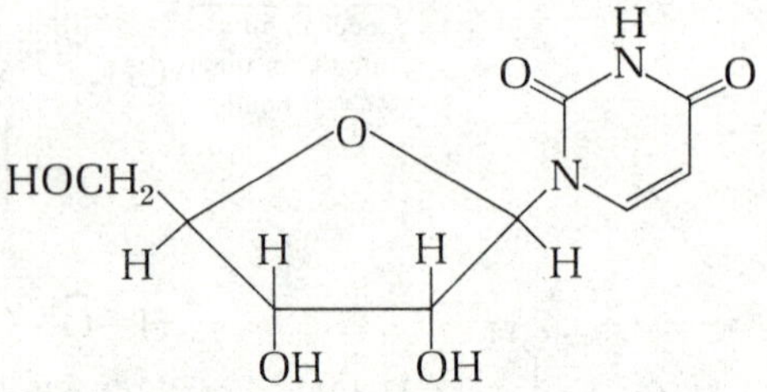

c) HOCH$_2$

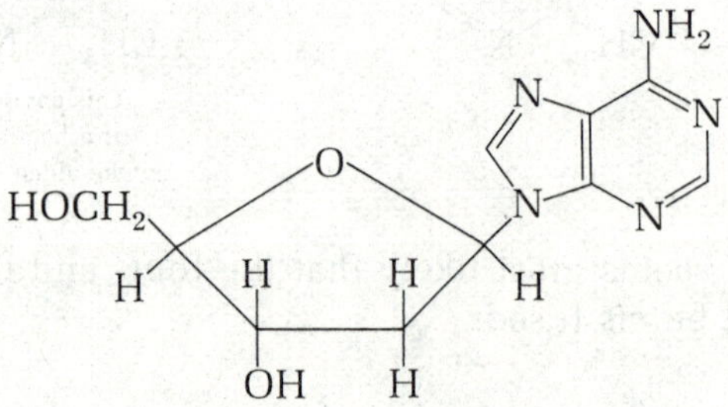

deoxyadenosine

d) HOCH₂

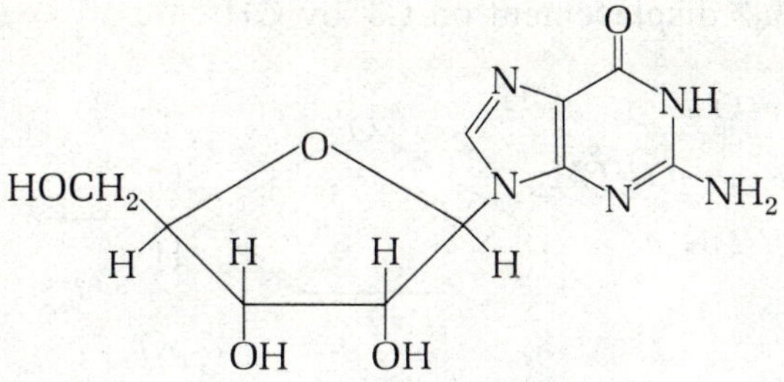

guanosine

e) HOCH₂

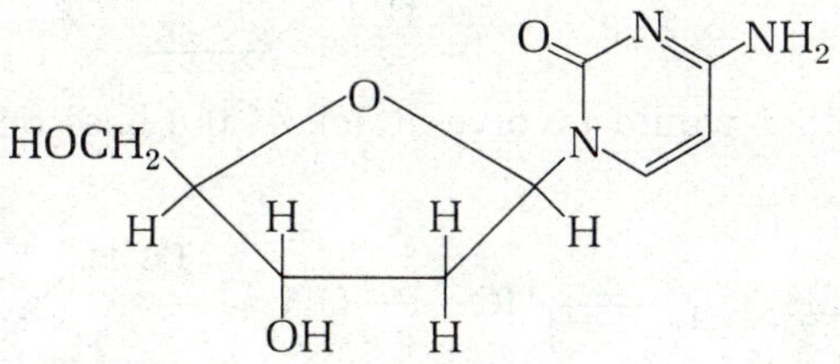

deoxycytidine

18-17. *a)* Only the essential parts of the formulas are shown.

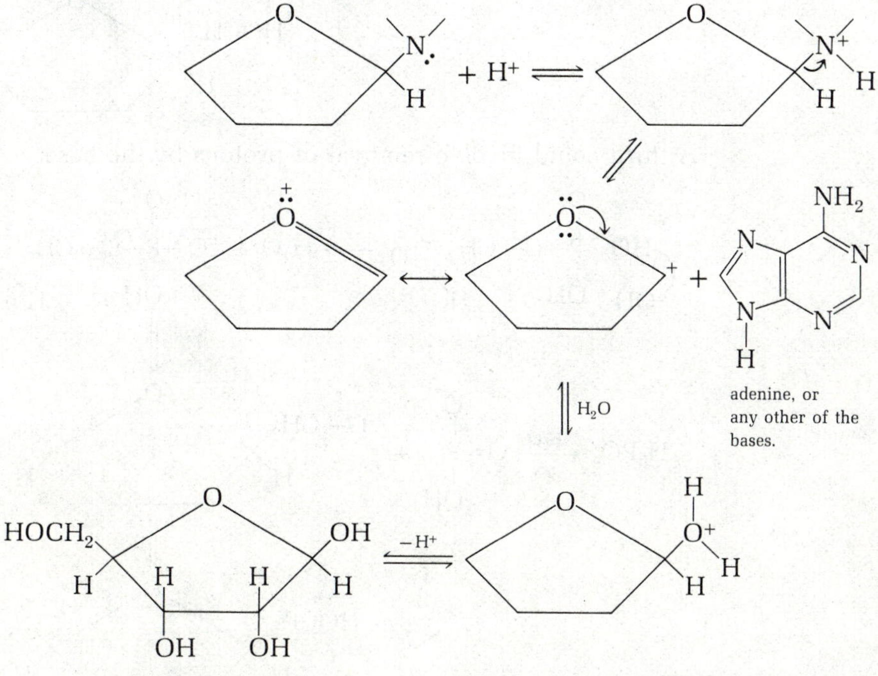

+ H⁺ ⇌

⟷ +

adenine, or
any other of the
bases.

H₂O

HOCH₂

ribose (or deoxyribose)

– H⁺ ⇌

b) At least three simple mechanisms are possible; one is a direct

S_N2 displacement on C5′ by OH⁻.

$$HO-\overset{\displaystyle O}{\underset{\displaystyle OH}{\overset{\|}{P}}}-O-CH_2 \quad\quad + \quad OH^- \longrightarrow$$

$$HOCH_2 \quad\quad + \quad HO-\overset{\displaystyle O}{\underset{\displaystyle OH}{\overset{\|}{P}}}-O^-$$

Another would involve attack of the base on phosphorus:

$$HO-\overset{\displaystyle O}{\underset{\displaystyle ^-OH\ \ OH}{\overset{\|}{P}}}-O-CH_2 \longrightarrow HO-\overset{\displaystyle O}{\underset{\displaystyle OH}{\overset{\|}{P}}}-OH \ + \ ^-OCH_2$$

$$\Big\Updownarrow H_2O$$

$$HOCH_2$$

A third could involve removal of protons by the base:

$$HO-\overset{\displaystyle O}{\underset{\displaystyle ^-OH\ \ OH}{\overset{\|}{P}}}-O-CH_2 \longrightarrow H_2O \ + \ ^-O-\overset{\displaystyle O}{\underset{\displaystyle OH}{\overset{\|}{P}}}-O-CH_2$$

$$\Big\Updownarrow$$

$$H_3PO_4 \xleftarrow{H_2O} O=\overset{\displaystyle}{\underset{\displaystyle OH}{P}} \ + \ ^-O-CH_2$$

$$\Big\Updownarrow H_2O$$

$$HOCH_2$$

More complex mechanisms, involving participation by nearby hydroxyl groups on the sugar can also be involved.

18-18.

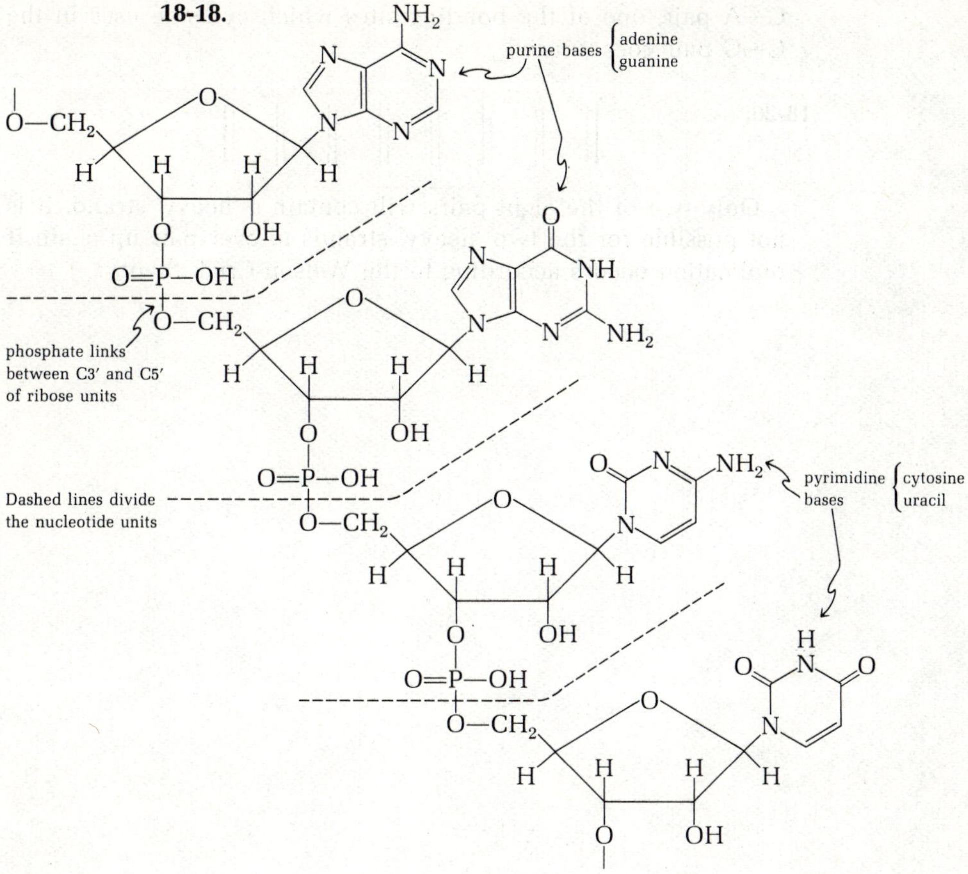

purine bases { adenine, guanine

phosphate links between C3′ and C5′ of ribose units

pyrimidine bases { cytosine, uracil

Dashed lines divide the nucleotide units

18-19. Possible structures are:

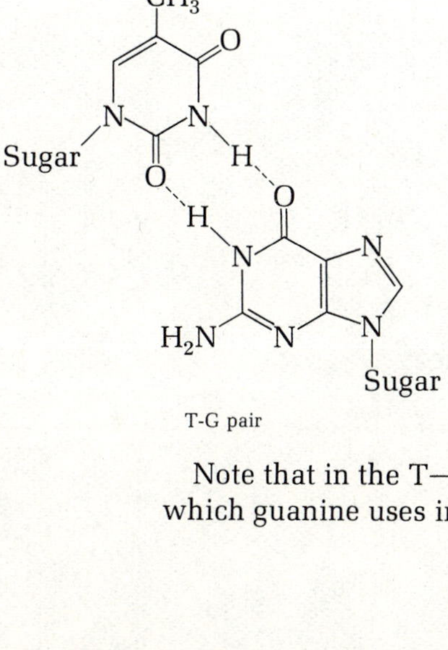

T-G pair

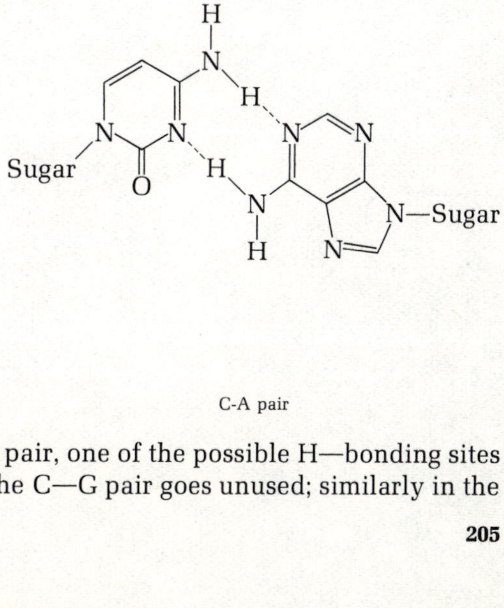

C-A pair

Note that in the T—G pair, one of the possible H—bonding sites which guanine uses in the C—G pair goes unused; similarly in the

C—A pair, one of the bonding sites which cytosine uses in the C—G pair goes unused.

18-20.

‖ ‖ ‖ ‖ ‖ ‖ ‖ ‖

Only two of the eight pairs will contain a 'heavy' strand. It is not possible for the two 'heavy' strands to ever pair up again if replication occurs according to the Watson-Crick theory.